Robert Campbell Auld

The Breed that Beats the Record

And Wins in the Race for Supremacy as the Most Economical Producer of the Primest Meat

for the Million

Robert Campbell Auld

The Breed that Beats the Record
And Wins in the Race for Supremacy as the Most Economical Producer of the Primest Meat for the Million

ISBN/EAN: 9783743423718

Manufactured in Europe, USA, Canada, Australia, Japa

Cover: Foto ©berggeist007 / pixelio.de

Manufactured and distributed by brebook publishing software (www.brebook.com)

Robert Campbell Auld

The Breed that Beats the Record

THE BREED
THAT BEATS THE RECORD

AND WINS IN THE RACE FOR SUPREMACY AS THE MOST ECONOMICAL

PRODUCER OF THE PRIMEST MEAT FOR THE MILLION.

A DEMONSTRATION

OF THE

PROPERTIES, PREPOTENCE, PRE-EMINENCE AND PRESTIGE,

OF

ABERDEEN-ANGUS

THE POLLED CATTLE.

WITH AN INTRODUCTION BY
JUDGE J. S. GOODWIN, A.M., BELOIT, KANSAS.

"The Polled Cattle compel the attention of the civilized world.'

DETROIT:
Aldine Company, 40 and 42 Congress Street West
1886.

CONTENTS.

Page

LIST OF ILLUSTRATIONS, . x

INTRODUCTION, . . xi

CHAPTER I.

NATIVE HABITAT AND PHYSICAL CONDITIONS.

Home of the breed—North-eastern counties—These in former and présent times—Geological formation—The Grampians—Granite—Lower down—Gneiss, mica-schist, etc., and red sandstone—Soil—Naturally infertile—By cultivation made very productive—"Timberless Buchan"—Soil in Forfarshire and "The Mearns"—Climate—Depending on heat and moisture, very wretched—Low mean temperature and range—At coast, inland, and among the mountains—Rainfall—Crops—Chiefly turnips, grass and oats—Barley, wheat, etc., only in favorable spots—The district described by a Frenchman, the author of "Le Betail en Ecosse"—"Very windy"—By an American—Mr. Wm. Warfield, the eminent Shorthorn authority—Graceful tribute to Aberdeenshire Shorthorns and "Prime Scots"—Comments on the conditions of rearing—Resulting in the production of the native Polled Aberdeen as the "Prime Scots—Described by a native and breeder of the rival roan—A remarkable article from the *Breeder's Gazette*—The Aberdeenshire Shorthorn founded on the cross with the native breed—Proof of the value of Aberdeen-Angus blood in giving *backbone*—Remarks of M. de la Trehonnais, well-known French authority on "Durhams," corrected—Testimony of Clement Stephenson, F.R.C.V.S., *the every year champion*, ... 1

CONTENTS.

CHAPTER II.

THE BUCOLIC IDEAL.

Mr. McCombie's description of his ideal animal—The author's summed up in "*as even from end to end as an egg*" and "*beef from the lug to the heel*"—The eye, head, and crown fully commented on—Color and pile—Standard color black—The hardiest color—So stated by M. Paul Marchal—Color of the breed in early times—Dun the color of the fairy cattle—Dr. Skene Keith and Mr. Headrick quoted—"Variegated with white universally disesteemed"—Pile early attended to—To withstand the rigors of winter—Two coats—An under and upper "overcoat"—Hide—Weighs lightest at Chicago—And sells highest in Aberdeen market—Connection with ancient *Urus* accounting for certain variations—The red color—Bowie's and Fullerton's opinions—"Purest and best"—Hon. M. H. Cochrane's, of Hillhurst, experiments —A "comely" couplet—"Black polls to fill the land like the black hogs"—A remarkable quotation from 1882 (London) *Rural Almanac*—"White points coincide with coarse fibre, and black with fine fibre and compact size"— Black the result of fancy—Origin of the term "black cattle"—Early preference for black cattle by Varro, etc.—By Markham in 16th, and Lawrence early in the present century—Samples—Mr. Hannay's "Lady Paramount," Ballindalloch "Judge"—Death of latter in possession of Judge J. S. Goodwin, Beloit, Kansas— 1879 Smithfield "reserve"—Clement Stephenson's 1884 champion—Comparison with the first scale of bovine points—Written 600 B. C., by Mago the Carthaginian— The standard for two thousand years—Up to the *muscular maturity* ideal of Prof. G. T. Brown's advocacy in his new book "Animal Life," 12

CHAPTER III.

THE MILKY WAY.

Dairy properties—The breed first famed for their dairy qualities—For which sought out "all over the North" in the beginning of the century—Large export of dairy produce from Aberdeenshire—Youat stated the Buchan cows were equal to the Ayrshires in quality. Records of Portlethen Herd in 1845—Quantity of milk given by show-yard animals—The late Lord Airlie's experience of the breed as fitted for the dairy—Highly satisfactory—Results in the late Fyvie Herd—Other numerous instances not necessary—"Buchan Prime in Buchan Rhyme"—Another remarkable cow—Baron de Fonteney's high opinion of the quality of the milk—General considerations—Beefing qualities were developed pre-eminently at expense of milking qualities—Three remarkable quotations from *Daily Free Press*, of Aberdeen, Scotland, ... 23

CHAPTER IV.

VIGOR.

Breeding qualities—Twins and triplets—A unique case of a calf-producing cow—25 calves in 7 years—18 came to maturity—Remarkable cases of vigor of famous cows—"Old Grannie," "Black Meg," etc.—And bulls—Constitution—T. W. Harvey's experience—Superior to the severe conditions of rearing—Statement by the *Farmer's Review*—"Can winter anywhere"—System of wintering in Aberdeenshire—Exposed to winter rigors—Opinions of Principal Walley, P.R.C.V.S., and Clement Stephenson, F.R.C.V.S., on the resistance of the breed to tubercular troubles—One of their chiefest advantages—An article from the London *Live Stock Journal*—Showing how in this the Polled Aberdeen-Angus "have the pull"—viz., in possession of "lean flesh"—the secret of their success in the show-yard and in disease resistance, . 39

CHAPTER V.

RANGERS.

Utility of want of horns—The London *World* and "Jim Lowther's" Polled Scots—Mr. Geo. Hendrie, of Detroit, likes them for the same reason—"They can't hook the colts"—L. A. Hine's experience of their docility--E. G. Underhill's—Feeding and grazing qualities—"Graze all over the world"—In Australia—New Zealand—On the American plains—Mr. James Macdonald's visit to the Victoria Ranch, Kansas — In Demerara — Jamaica — Buenos Ayres—Gov. O. A. Hoadley's experience of the breed as rustlers in New Mexico — President T. W. Holt's—W. H. Embry's—Judge J. S. Goodwin quotes two experiences of the West—*The Rocky Mountain Husbandman's* statement of the success of the breed in Montana—The experience of the Montana Cattle Company—Messrs. Martin & Myers—*The North-Western Live Stock Journal*--Cattle supplied by Messrs. Estill & Elliot "Rustling"—Success of the breed recorded in *The Field* —The Cochrane Ranch, Bow River—Mr. Geo. Findlay's interviews with ranchmen—Mr. J. J. Hill's testimony—*National Live Stock Journal*—The breed "answers admirably in Texas"—Stands the alternations in climate—Come out better than any other imported cattle—S. P. Cunningham makes similar statement—*The American Agriculturist* and its pictorial proof of superiority of the breed for the West, ... 38

CHAPTER VI.

GRADES.

Crosses and crossing—Numerous proofs of the superiority "over all" of the Shorthorn-Aberdeen cross—The influence of the Aberdeenshire dam—Convincing proofs of this—The Polled sire preferred—Makes a cross superior to the former—The best in the world—Proofs—The Aberdeen sire used everywhere--In Galloway still—All over Scotland—Ireland—Every day more so—Statements of the late H. D. Adamson—Mr. Hine—Mr. T. W. Harvey *The Canadian Live Stock Journal* and Aberdeen grades—Experience of Geary Bros.--Details of Hon. M. H. Cochrane's experience—Hon. Mr. Pope—J. J. Rodgers—Estill & Elliot—Leonard Bros.—The verdict—Crosses with dairy breeds—Ayrshire—Holstein—Jersey, . . 48

CONTENTS.

CHAPTER VII.

EARLY MATURITY.

Size—Authority of James Bruce, Ruthwell, Annan, Scotland—T. W. Harvey, Chicago, Ill.—Hon. F. Allen—Jas. Macdonald—Late H. D. Adamson—Robert Bruce, Great Smeaton, Northallerton, England —Weights —Offals— Bone of Watson's Smithfield heifer *like that of a deer*— Light offal—Proved by high per cent. of net to gross— Instances of light bone by Charles Bruce, cattle salesman, Newcastle — Important extract from *Scotsman* — Dress over 72 per cent. at Poissy, 1857—Early maturity— Henry Evershed's statement—In 1881 Aberdeen youngsters " sweep the boards " at Smithfield--Gain per day at Smithfield, 1879-1880—Not *beef* that always makes weight —In 1881--Tables—*The Field* on the Altyre champions— Comparison with the other breeds—" Aged steers " once a hobby of the older breeders—New order of things— —Birmingham averages of 1883—Winners of " Breed Cups" at Smithfield, 1883—Remarkable figures of the breed in 1885, ... 61

CHAPTER VIII.

PRIME SCOTS.

As beef producers—Aberdeen beef famed in 4th century— And ever since—How described in 1810—Dr. A. Forsyth states in 1810 " they easily top the market at London "— Then fed on plain turnips and straw—Since then the term Prime Scots has solely applied to the Aberdeen Polled cattle—The *Farmer's Review* on this—*Mark Lane Express*—" An Aberdeen is thicker than a Hereford or Shorthorn "— *Country Gentleman* (London) on " Prime Aberdeens "- -Mr. Hine visits London market—Col. G. W. Henry—G. L. Stietcher—Special reports in all journals —John Chalmers Morton, editor of *The Agricultural Gazette*, etc.—London *Daily Telegraph* on the Christmas market—H. D. Adamson—Mr. J. L. Thompson, of South Australia— An example report — Comments — Mr. Wm. Anderson's (Wellhouse), experience of Prime Scots—

Quarterly Review, 1857—London *Standard—Times—Mark Lane Express*—Mr. G. T. Turner, *Live Stock Journal*—Mr. McCombie's memorable remarks at Rothiemay sale—"Autobiography of a 'Prime Scot,'" by Druid, in *All the Year Round*—*Punch* on Polled Scots, . . 72

CHAPTER IX.

THE EPICURE'S CHOICE.

The beef—*National Live Stock Journal* on the beef—*Le Fermier*—Beef recherches—*The Agricultural Gazette*, and test of Smithfield Club—At Chicago—Representation of Polled Aberdeen beef—Remarks of Aberdeen *Free Press*—Of *Breeder's Gazette*—Mr. T. W. Harvey on the beef—Geary Brothers and the meat of their steer "Black Prince"—At Kansas City—Beef of Col. Henry's "Bride 3rd"—Mr. P. D. Armour's compliment to the beef—Mr. J. J. Hill's steer "Benholm"—Croxson's (Stock Yards restaurant) choice—G. T. Williams, of the Stock Yards, declared "no man alive ever tasted better beef"—Mr. John B. Sherman again passes high encomiums on Angus beef, . 83

CHAPTER X.

TUNING UP.

The Polls in American Fat Stock Shows—Sparingly exhibited—1883, KANSAS CITY—Geary's "Black Prince"—Henry's "Bride"—Gudgell & Simpson's—"Bruce's Queen"—Matthew's "Paris Heifer"—CHICAGO—Cochrane's "Waterside Jock"—Dressed the largest per cent. of net to gross of any two-year-old of any year; report of committee on him—Cochrane's cow, "Duchess 2nd, first in class of cows open to all; committee's report—"Black Prince," breed champion, also sweepstakes in class judged by butchers—Flattering report of committee—1884, KANSAS CITY—Grade "Abernethy," second in two-year-old grade class, first in class for *cost of production*—"Blaine" and "Logan" (Indiana Company) first and second in yearling classes for *early maturity*, and *cost of production*—Henry's "Bride 3rd" of Blairshinnoch—Makes a clean sweep of everything—A remarkable series of victories for "a half-fed Angus"—H. D. Adamson's report of the show—"Burleigh's Pride"—A cross-bred Aberdeen-Hereford—Gains *Breeder's Gazette* challenge shield—A living monument to the influence of Angus sire—Characterised as a *very remarkable animal*—Purchased by Dr. C. J. Alloway, of Grand Forks, Dakota—The Angus sire has the merit—CHICAGO—Cochrane's "Netherwood Jack;" breed sweepstakes—"Blaine" "Logan"—"Abernethy," grade—"Quality," grade—The last of "Black Prince"—Weight, 2,600 lbs.—Dressed over 71.50 per cent.—Success achieved by polls at shows—A special quotation for Prime Scots in America—Aberdeen-Angus grow as much as any during first year—1885, KANSAS CITY—Gudgell & Simpson's "Sandy" takes the breed sweepstakes, and The Polled Cattle Society's gold medal—"One of the best doddies ever seen"—"Uncle William Watson'—Grades and crosses—CHICAGO—"Benholm" takes breed championship and The Polled Cattle Society's gold medal—"Wildy"—"Blaine" and "Logan"—Committee's report—*Gazette's* report—Consolation class, "Benholm" and "Sandy" champions—"Benholm" highest net to gross on record in America—Grades—*Gazette's* report—Comparison with other breeds favorable to Aberdeen-Angus—*Farmer's Review* and edible beef, "the aim of the exhibitor"—"Turriff" should have carried carcass sweepstakes—Exhibit at American shows "small, but select"—The breed destined to "Clement-Stephenson" all other breeds, ... 90

CHAPTER XI.

APOTHEOSIS.

PART I.—The colossal victories of the breed for the last 30 years—THE CENTENARY SHOW OF THE HIGHLAND SOCIETY, 1884—Official report of the Society on the Polled Aberdeen-Angus Cattle, written by Rev. John Gillespie, A. M., Mouswald—Justice, Prince Albert of Baads, The Black Knight, Waterside, Matilda 2d, Electra—Justice heads the Ballindalloch Prize Group, and Matilda the Waterside Prize Family—Breeding of Justice—His importation to this country—Breeding of the other Prize-winners—Proof that "blood always tells"—Editorial comments of London *Live-Stock Journal*, highly flattering, as usual, to the "black-skins"—Mr. Geo. Hendry of the *Daily Free Press*, publishes valuable testimony—PART II.—BIRMINGHAM AND SMITHFIELD, in 1885—Extracts from reports of correspondents of the various stock organs of Britain and America on the appearance of the breed—"Makes a marked sensation"—"The best at the exhibition"—"The chief honors fall to them"—"LUXURY," the double champion at both places—"The breed on the highest pinnacle of fame"—The cow class "not excelled by any in the exhibitions"—Aberdeen-Angus "black is the dominent color, for beef"—"Accomplished a feat unparalleled in fat-stock shows"—"No one breed had such types as had the black polls"—The Altyre exhibits—"Next" to Luxury—Altyre should have been "Reserve" of the Smithfield—Aberdeen Polls the "*Scots*" of the early Norfolk graziers—The scene during the award of the Smithfield "blue ribbon"—"*Heather Bloom! Heather Bloom!*" once more. Winnings of "Luxury"—Her sale, 2s. 6d. per lb.—Particulars of her slaughter, carcass, etc,—Dresses 76 per cent. net to gross—"*Prodigious!*"—Breeding of Luxury, 105

CHAPTER XII.

APOTHEOSIS—CONTINUED.

PART III.—The Aberdeen crosses—The cross-breds by the universal acclamation of the press, declared to have been the best part of the shows—The chief prize-winners were sired by Aberdeen Polled bulls—Effects of this cross—*The Field* declares this cross "the most successful ever tried—*The best in the world*—Analysis of the cross-bred classes—Breeding, prizes won, character, comments and comparisons gathered from many a flower in the gardens of live-stock journalism—The records is of Aberdeen-Angus only—The Galloway makes no fight—The "difference" between the Galloways and the Aberdeen-Angus—As stated by Mr. Gordon, chief in-

spector of stock for Queensland, and others—PART IV.
—The London and Christmas markets, 1885—Shows
what is meant there by such terms as "Scots," "Polled
Scots," "Black Polls"—For quality the best market in
the world—Reports of all the live-stock journals stating
that the Prime Scots are solely, wholely and entirely
Polled Aberdeen-Angus, . - - . 127

ADDENDA.

A London Polled Aberdeen Society—The Breed extending
its outposts far and near—Replacing other Breeds—
Converts and adherents from rival breeds—Clement
Stephenson—Hon. M. H. Cochrane—G. W. Henry—
The *Field* and Judge Goodwin—"Comparative" tests of
Agricultural Colleges exposed—Test of public sales—
Sales of 1883, 1884, 1885, in favor of Angus—The last sale,
Hon.M.H.Cochrane leads them all—The Aberdeen draws
the premium at a dairy contest—Crossing—Polled bulls
bought largely in Britain and Ireland for crossing on the
other breeds—Mr.J.A.Cochrane's (Hillhurst) report—J.S.
Goodwin—Testimony of high British authorities—Early
maturity, by G. T. Turner, again—Chicago "stock yards"
and other dealer's and butcher's opinions as to the breed
"on foot" and "on block"—In the West—The descendants
of the Victoria Ranch Aberdeen Polled bulls—Mr. Wm.
Watson, of Keillor, on the breed in the West—The
"Dalmore" herd for New Mexico—This new herd a
special feather in the Aberdeen Angus cap—The purchase of Gavenwood and Glenbarry herd by the Geary
Bros., of Guelph, Ont.—Mr. R. C. Dye captivated by the
Aberdeens, prefers them to his former love, the Jerseys,
—CONCLUSION—Prof. Brown, Ont., and the breed the
world's new beefer—The Aberdeen "the coming steer"
—*Drover's Journal* and *Farmer's Review* on horns and
dehorning—Horns not useful, and not ornamental—
Hornless character says the *Nationol Live-Stock Journal*:
"the least of the good points of the Angus"—The breed
"morally certain to win—"Competitors beware!"—
American Agriculturist declaration in favor of the breed—
"Have won the confidence of America" and "the best
breed of beef cattle in the world"—The grand general

summing up of the American Fat Stock Show official report—Places the Aberdeen first, their grades second and "the field" to follow—" Lastly," *They are the chunky sort that means business*—BIBLIOGRAPHY—THE AMERICAN ABERDEEN-ANGUS ASSOCIATION—ERRATA, . . 145

ILLUSTRATIONS.

1. ALEX. RAMSAY, ESQ., (From photograph by Moffat, Edinburgh, Scotland), . . . Frontispiece
2. CROSS-BRED CHAMPION AT BIRMINGHAM, 1884, (From London *Live-Stock Journal*), ... 53
3. ABERDEEN BEEF, by A. M. Williams, London, (From Photograph by Wilson, Aberdeen, Scot.) . . 85
4. JUSTICE, (From London *Live-Stock Journal*), . . 109
5. LUXURY, Birmingham and Smithfield Champion, 1885. (After London *Live-Stock Journal*.) 114
6. POLLED HEAD, BASUTO, Tailpiece, 142

INTRODUCTION.

To conform to the usages of polite society, this book may need an introduction, but the Angus Doddie introduces herself. No one who has seen the beautiful hornless head, intelligent eye, symmetrical form and velvety coat of an Angus Doddie, has ever waited for any formality before pressing to a closer acquaintance.

The past decade has brought laurels enough to the polled head to warrant the statement that the Aberdeen-Angus is, indeed, "THE BREED THAT BEATS THE RECORD." From that day, in 1878, when Paris and his herd, and Judge, the "World Beater," carried away the highest honors at the Universal Exposition in Paris, down to the triumphant show-yard career of Luxury, in 1885, the star of Angus supremacy has never dimmed. They are no longer an "experiment" in America, but hold a front rank among the come-to-stay breeds.

There is room for all improved breeds. Each one will have is warm supporters and earnest advocates. In the contest for supremacy, it is *facts*, not *talk*, that counts, and this work offers an array of facts that cannot be controverted or passed over in silence. It establishes the following assertions:

1st. That Angus cattle weigh as heavy as any other breed.
2nd. That they mature as early.
3rd. That they dress a larger per cent. of dead to live weight than any other breed
4th. That they are a strong, hardy, vigorous race of cattle.
5th. That they are as good, if not better, milkers than any other beef breed.

INTRODUCTION.

6th. That, being polled, they are easier to handle, do less damage, require less room and consequently less money in handling.

7th. That they are unsurpassed by any other breed for "crossing" or "grading up."

8th. That they, alone, are the PRIME SCOTS of the British and other market quotations.

These facts alone show that they are worthy of the pride and esteem in which they are held by their admirers, and should be enough to command careful attention to the following pages.

To the knowing ones it is enough to say in conclusion, that the Scotch cattle are as good and true as Scotch hospitality, and more than that pen cannot write.

JOHN S. GOODWIN,
Beloit, Kansas.

The Breed that Beats the Record.

CHAPTER I.

Native Habitat and Physical Conditions.

HOME OF THE BREED.

It will be sufficient for the present purposes of this work to confine our view to that group of counties generally, collectively called the north-eastern counties of Scotland. These are Forfarshire, formerly Angusshire; Kincardineshire (the southern part of which is " *The Mearns*," the northern belonging to Mar); Aberdeenshire, comprehending most of Mar and Buchan, which two districts have several sub-divisions, such as Strathbogie, Formartin, Garioch, Alford, etc.; Banffshire, which formerly was included in Buchan; Moray and Nairn shires. Of these, Aberdeen is, of course, by far the largest, comprehending eighty-five parishes; Forfar has fifty-five; Kincardine twenty; Banff twenty-five; Elgin and Moray twenty-seven.

Aberdeenshire is so large that this should be borne in mind by those studying the history of the breed in that county. This county consists of wide "low-land"

and "high-land" regions—Buchan constituting nearly all the former, with twenty-four parishes, Mar comprehending the latter, and also including the parishes of Kincardineshire, drained by the Dee, and on the north side of the Grampians; so that Aberdeenshire is constituted to-day by the chief parts of two ancient provinces, the wings of which have been separated from it.

GEOLOGICAL FORMATION.

The Grampian range forms the controlling physical feature of this region, running north to opposite Stonehaven and forming the Highland region within its irregular boundary. The Grampians consist of upheaved granite—this distinguishing primitive formation coming there to the surface. Lower down, overlaying the granite, occur gneiss, mica-schist, quartz, limestone and clay-slate. Of such rocks does the greater part of the cultivated region of Aberdeenshire consist. In Kincardineshire, Forfarshire and Morayshire the old red sandstone exists in the "laichs."

SOIL.

The character of the soil varies. In Aberdeenshire, excepting in certain alluvial tracts, it is not naturally rich or fertile, consisting for the most part of a thin coating of vegetable mould resting on the coarse glacial or blue-clay. But by the most perfect cultivation the soil has been brought to the highest pitch of productiveness, and grows great crops of turnips, oats and grass. Buchan is timberless; but most of it was fertile in the earliest times, and so was Garioch. The former was the "granary" and the latter was the "girnal" of the north. Morayshire was also named the Garden of Scotland. The soil in the "laich" of Moray is particularly responsive. In Mearns and Forfarshire the soil is red clay, gravelly loam and "carse."

CLIMATE.

This depends on heat and moisture—these, as need hardly be said, being in the northeast of Scotland, low and great. This means a "wretched" climate. The mean temperature is low, from 40 degrees to 47 degrees, with a range of 10 degrees to 20 degrees above or below, for summer or winter. Proximity to the coast modifies the climate; while distance inland from it, combined with altitude, tends greatly to intensify its severity. The rainfall is affected by the winds, which seem mostly to conspire to blow off the coast. The winds from the southwest, borne from the Atlantic, part with their moisture over the Grampians, before reaching the heart of Aberdeenshire. But the east and southeast winds, always cold and biting, are impregnated with moisture. The north winds are cold but dry, and are a God-send in a late, damp harvest—the great draw-back to Aberdeenshire agriculture.

Rainfall varies with locality; among the southwest mountains it is stated at 100 inches per annum; in the middle of the country, 40 inches; in the valleys of the Dee and Don, and northeast lowland, 30 inches.

CROPS.

These are chiefly those needed in cattle feeding— yellow turnips and swedes; grass, Italian rye, etc., mixed with red and white clover—the latter growing almost naturally—and oats. Barley is grown very moderately, wheat hardly at all, in Aberdeenshire.

In Moray, "The Mearns" and Forfar, wheat, barley, potatoes, beans and peas are successfully cultivated. But in the cattle regions proper the chief crops are those first stated. The granite soil, indeed, is particularly adapted to grow these nutritiously.

THE DISTRICT AS DESCRIBED BY A FOREIGNER.

Baron L. de Fontenay, author of "Le Betail en Eccossais," gives us an idea of what a foreigner thinks of the middle district of Aberdeenshire, where he was resident, as a student, for some years. "The country is moist and very windy. As to the temperature I cannot speak with precision, there having been absolutely no winter in 1858-9 in Scotland any more than in France. But I believe the thermometer does not fall so low as in Normandy. I found myself in the midst of mountains, the country in this direction being only a succession of hills and valleys. The valleys and hillsides are cultivated; the mountains are in part covered with Scots fir; their summits are barren and present on the surface only a short heath, often incapable of producing the least pasture for sheep. The soil is granitic and schistoic; in some places are found soils entirely analogous to those of Bretagne."

BY AN AMERICAN.

In describing the influences of this climate upon shorthorns, Mr. Wm. Warfield says: "Away up in

Aberdeenshire, exposed to all the rigor of that extreme north, this herd (Mr. Amos Cruickshank's) has been for many years a grand example of what shorthorns may become in Scotland. I have never seen finer fleshed, larger framed, richer coated beasts anywhere than this herd. Scotland's eminence, as a beef producing country, is too well known to need any particular comment, and 'Prime Scots,' as the top quotation in English markets, is an old story. There is something singularly taking in the whole class of Scotch cattle. What blocky, low-down beasts they are! You will be told anywhere in England, by feeders, that Scotch bred and fed cattle will go to the block in better form than any south of the border." Mr. Warfield also alludes to "the great capacity of all classes of animals bred in cold climates, to make peculiarly rapid and vigorous growth during the summer, a capacity shared by all nature, and the tendency to lay up fat, as if stored for the long winter's drain on the system. The effect of the bracing, invigorating air on the whole constitution, deepening the chest, filling out the form in every way needed to baffle the winter's cold. Springing from these, we find an active digestion, rapid assimilation and fine flesh producing qualities." These are the conditions of climate, in which the Polled Aberdeen-Angus have been reared from time immemorial, and, as a result, we have this well defined type of cattle, having great substance, great aptitude to fatten, and of early maturity and hardiness; to all of which reference will again be made. It is these native polls, remember, that constitute the *prime* of the "prime Scots" that have been regarded so long with such solicitous affection by Mr. Warfield and other eminent shorthorn men.

THE BREED THAT BEATS THE RECORD.

AS DESCRIBED BY A NATIVE, AND BREEDER OF THE RIVAL ROAN.

In the Breeder's Gazette, September 10th, 1885, appeared an article "From far-off Aberdeen—A Word from the Land of prime Butchers' Beasts," by the well known breeder of Aberdeenshire shorthorns. We are tempted to give it in full. The writer goes on the principle of the British newspaper editor, who ignores his esteemed contemporaries altogether, proceeding in his course as if *they were not*. The article is, however, so interesting and so full of point that by substituting here and there the words *Polled Aberdeen* for *Shorthorn* one will get a tolerably vivid realization of the natural conditions out of which the prime Aberdeen has arisen, like Aphrodite from the sea foam. If it has been so with the *imported* article—the shorthorn—how much more so must it be with the *native* Aberdeen, that, mind, has had no contamination with "the poison" —as Aberdeen breeders regarded it—of the universal intruder :

Sixty years ago the Northeast of Scotland grew little more either of beef or corn than was wanted for the comparatively scant population. The use of lime and bones and the draining of the land opened the way for turnip culture, and now what was one of the poorest districts of the whole kingdom is one of the principal sources of meat supply of the finest quality. This has been brought about by the Shorthorn cross on the native breeds of cattle, and hence it has come to pass that the particular variety of Shorthorn known as the Aberdeenshire Shorthorn has special interest for those who are opening out a new meat supply for the world.

The breeders of Shorthorns in Aberdeenshire have been mainly men who had to pay a rent and make a living by their own exertions. They have not been theorists; circumstances have not been favorable to the formation of opinions in favor of line breeding or any other of the modern improvements of which so much was heard some ten years ago. The necessity of keeping a house over his head has prevented the Aberdeenshire breeder from following the caprices of fashion; his customers have been men who were obliged to apply ruthlessly the test of utility. If a sire proved a bad one he must go, no matter how grandly his pedigree might look on paper; the calves would not sell, for they were to be used in producing beef, and if they could not do that the blood availed nothing. The blue-blooded weed for which there used to be a kindness in some directions was dreaded beyond everything; the very blueness of his blood made him the more dangerous.

Aberdeenshire is a cold, bleak country; there is very little timber, fences are formed of stone walls, and there is a chilliness and rawness in the east wind which is more trying than the cold of the western plains. The soil is, generally speaking, poor; indeed, no part of the world where improved cattle thrive is it so poor. No sort of life which is not unusually vigorous will survive, and hence the cattle bred in such a country must be hardy. The weed, blue-blooded or otherwise, is improved out of existence by the hard conditions of its life. Constitution has thus become the first necessity with the Aberdeenshire breeder, and as all other parts of the globe where cattle live are more suitable to animal life the Aberdeenshire Shorthorn takes kindly to exile and is the best colonist of his kind.

The making of beef has been the trade of the district. What every farmer wanted was a heavily-fleshed sire, and thus next to constitution the demand has been for this type of animal which has most aptitude for turning its food rapidly into beef. But besides all this, the farmer as a man of business had to look to the amount of the turnover of his capital and the quickness with which it could be effected. Two generations ago a man was satisfied to feed off an ox at four years old; twenty-five years ago many thought they did well to finish their cattle at three years; now all aim at feeding off at from twenty to thirty months old. Thus a demand for early maturity arose. The stamp of animal which looks shabby at two years but grows into a fair cow at six or seven is of no use; our cattle must be useful both when young and old. And as the three leading characteristics of the Aberdeenshire Shorthorn are nothing more than the outcome of the necessity of the district and its inhabitants, we must have (1) constitution, (2) a tendency to carry a great weight of beef, and (3) a capability of early maturity, otherwise we can neither pay our rents nor make a living.

It will be observed that nothing has been said about style, completeness of form, and purity of blood. To take the last point first, no one believes more firmly in the value of pedigree and purity of blood than the Aberdeenshire breeder. But he does not judge of the value of ancestry by its limitation to any particular strain; he values a pedigree in proportion to the known excellence of its representatives, and this excellence must be an existing fact of to-day, and not a matter of history or tradition more or less ancient. No tribe can live on its laurels in Aberdeenshire unless they are freshly won, but when a race of cattle can show a long record—not of show-yard successes—but of practical usefulness, and when that record is known to be still a growing one, the Aberdeenshire breeder will yield to no one in his appreciation of its value. Nor is he careless as to form and style; but here again the form and style which is valued is that which indicates the possession of the practical points of beef and milk producing. Well-sprung ribs and a broad chest are necessities for health; deep thighs; a well filled twist, a thickly covered loin, and finely formed bones, are required by the butcher, and as all farmers know that a good cow must be a good milker, the udder and milk vein are never overlooked without serious loss of usefulness and reputation.

The necessity of economy and the conviction that natural conditions are the most healthy prevent the use of any artificial food among matured cattle. Turnips and oat straw in winter and grass in summer suffice to keep the animals in health. The calves generally suck their dams, and a considerable quantity of milk is taken from the cows besides.

The animals which are the outcome of the severe conditions under which everything lives in this harsh country have proven themselves pioneers; in the States, in Canada, and in South America they have been tried. In their old home they have been hardy, vigorous rent-payers, and in their new homes they have already been equally successful.

HOW THESE ABERDEENSHIRE SHORTHORNS CAME INTO EXISTENCE.

Perhaps the above was written anent the following item in the same paper, headed "Aberdeenshire Shorthorns:"

THE BREED THAT BEATS THE RECORD.   9

"A writer in one of our London exchanges (evidently an Angus breeder) makes the following allusion to the practice followed by the early breeders of Shorthorns in Aberdeenshire which may be of interest to those of our readers who admire the Scotch cattle. He says: 'The Aberdeenshire breeders have indeed treated the shorthorn as they did the native poll in the early days—put that and that together, which they, in their accurate judgment, foresaw would not only nick, but produce the best stamp of beast for sacrificing at the shambles. If the local native poll did not help to mould them, they had their type reflected in them to a degree. Mr. Cruickshank began his breeding career on the native poll, but soon gave his entire allegiance to the Teeswater, and has been the chief breeder to originate the Aberdeenshire shorthorns.'" The eminent breeder just previously quoted indicates as much. All this is but a grand testimony to the backbone-giving character of Aberdeen-Angus "blood."

FROM THE FOREGOING AN ERROR CORRECTED.

Like the natives of the *genus homo* of the same region who have proven themselves such excellent colonizers, "pioneers," so have the native bovines most pre-eminently; so that such remarks as the following by that well known French authority, M. de la Trehonnais, require comment: "It is true that the Angus breed, notwithstanding its grand qualities from a beef point of view, is little known and less cared for in France. For my part, I do not know more than one breeder who would have been eccentric enough to have attempted to constitute and establish a herd of Angus cattle in our midst under conditions of soil and climate which are as adverse as can be imagined to the temperament and requirements of the breed. Angus cattle are not cosmopolitan, as shorthorns are. They have climateric, hygienic, and alimentary peculiarities which belong to them exclusively. It is on this account that although the breed is highly esteemed in England, as furnishing excellent meat, no breeder outside Scotland tries to acclimatise it; and even in Scotland is is only in the county of Aberdeen that the Angus breed is found [!] Few breeds amongst cattle are more exclusively localised than is this one."

Thus has a foreigner learned of the cattle that have become world famous!—Localized! Why! after the shorthorn it is the most " cosmopolitan" breed in Britain, as all the show-yards in the three kingdoms testify. And still they do not miss the climateric, hygienic and alimentary peculiarities of Aberdeenshire. Neither are the peculiarities of climate, housing or feeding of their native place missed in America. This is the old foible that we used to hear about its being the turnips and straw, the climate and soil, in fact the air and the water too, that have made the prime Polled Scots in Aberdeenshire. Look at them in rich England, grassy Ireland, and corn and prairie America—as the sequel will abundantly prove!

HEAR WHAT MR. STEPHENSON HAS TO SAY.

This now very eminent breeder and exhibitor of Aberdeens, says: " Some of my friends questioned if these cattle would do as well in this part of the country as they did in the North of Scotland, while others affirmed, quite confidently, that they were the breed of a particular locality, and would not suit this district. After what every one must admit has been a fair trial, it affords me more and more pleasure to show my herd to judges of stock; and every one who has seen them will bear me out when I say that in Northumberland we can grow them as big, if not bigger, and breed them as truly, as they can do in the North of Scotland."

CHAPTER II.

The Bucolic Ideal.

THE OUTWARD FEATURES OF THE TYPICAL ANIMAL.

Mr. McCombie, in his "Cattle and Cattle Breeders," thus describes his ideal animal: " A perfect breeding or feeding animal should have a fine expression of countenance. I could point it out, but it is difficult to describe on paper. It should be mild, serene and expressive. The animal should be fine in the bone, with clean muzzle, a tail like a rat; and not ewe-necked ; short on the legs. He should have a small, well put on head, prominent eye, a skin not too thick nor too thin ; should be covered with silky hair to the touch like a lady's glove; should have a good belly to hold his meat ; should be straight backed, well ribbed up and well ribbed home ; his hook bones should not be too wide apart. The wide hooked animal, especially a cow after calving, always has a vacancy between the hook bones and the tail, and a want of the most valuable part of the carcass. I detest to see hook bones too wide apart ; they should correspond with the other proportions of the body. A level line should run from hook to tail. He should be well set on at the tail, free of patchiness there and all over, with deep thighs, that the butcher may get his second round and prominent brisket; deep in the fore rib, with a good purse below him, which is always worth £1 to him in the London market; well fleshed in the fore-breast, with equal covering of fine flesh all over his carcass, so valuable to the butcher. His outline ought to be such that if a tape is stretched from the fore-shoulder to the thigh and from the back to the extremity, there the line should lie close, with no vacancies, and without a void; the line should fill from the hook to the tail. From the shoulder-blade to the head should be well filled up, as we say, good in the neck vein. These remarks as to the quality and proportions a beast should possess are set down, and not in good order, just as they struck me at the time. Thick legs, thick skin and bristly hair, always point to sluggish feeders."

Looking at the animal broadside, from the poll and crops to the tail, should be "straight as a rash," the neck in the female should only rise *if at all*, very slightly from the top-line level, in the male, with a more prominently rounded crest. In fact, we like to see the head projecting out on a direct level with the back in both sexes. The underline should be straight and parallel to the top line, with a prominent angled, keel-like brisket. In the female a neat umbilicus and full milk vessel are attractive. The tail-head should be neatly packed away; and there should appear no daylight between the tail and the buttocks—the tail should hang close. The neck should be small, with such an appearance that there is "none at all;" clean head, carved and bloody looking, with no coarseness. There should be no throatiness or dewlap. Such, indeed, reminds one of the old Italian-Spanish, or Zebu races, and indicates slow feeding qualities. Bone should be fine and smooth, limbs short and clean ; feet small. Viewed from behind the animal should look as if he was a complete cylinder, with no vacancy, but a barrel from end to end ; the rounded spring from the spine, enclosing, in an evenly covered contour, all the parts of the body. There should be no prominences, but the surface smoothed off all over, behind and before the hooks, and behind and before the shoulder, the line should have no hollow—the same from flank and fore flank. In such an animal the depth and width through buttock and heart will satisfy the most exacting. Everywhere, well filled all over, no gaudiness, but " as even from end to end as an egg."

When the hand is applied to the body it will come in contact with a well covered frame—no hard or sharp bones should be felt. The hand will sink evenly inward and move freely as if on the softest cushion. The pile will feel soft and skin and under skin will impalpably come together in the fingers and retire. The palm will float the superficial integument over the deeper covering. This magical touch indicates quality forever. " Beef from the lug to the heel "—tallowed within, with not an ounce of waste anywhere, the perfect beast stands on the shortest of legs before you.

The eye should be projecting, bright, quietly active and observant. It was such an eye as the gods gave Juno. It should be very full and its surrounding orbit should be prominently cushioned in the head. The lids wide, with no bushy eyelashes, but these long and fine, not obscuring the visual organs. Between the eyes the forehead should be broad, but from this upward it should run to a high peak or poll, which should have a tuft (sometimes " bald " with rubbing) and a distinct pit behind that will hold the fore fiingers. Everywhere the head should be clean and chilselled and the skin smooth and glistening. The ears should be well set on—the broad rounded "fan" rising above the neck where they are laid tightly back in their natural docile position. They should not, in the high-blooded, droop—even though lop-earedness might be thought a humel-doddie feature. The rounded fan should be hairy, the inside of a bright, waxy orange. The muzzle large but not coarse, dewy-healthy, the under chin prominent, a delicate pink nostril, and "sweet breathed."

LOOK AT THE POLLED CROWN CAREFULLY.

Such a head as we have described—and which is the natural one—will not exhibit any trace of scurs. These are no evidence of impurity: they are, like beauty, skin deep. It is needless to impress every breeder with the necessity of not using males with the least symptom of these, to any pure bred female. In fact, every male showing such symptoms should be steered.

COLOR AND PILE.

The standard color is black. This is the simple result of a fancy of the early improvers, hence it became a fashion and that rules rigidly and fantastically.

By the early breeders it was thought that black was the healthiest and hardiest color—that white or light shades were more delicate. They seemed to have some authority for this. M. Paul Marchal in *Revue Scientifique*, thus writes: " The intensity of coloration is generally proportioned to vital activity. * * * *

Breeders prefer animals rich in pigment matter, because they will resist disease, and most easily accommodate themselves to special systems of feeding. The ancients regarded animals having white hair on a black skin as the most vigorous. White parts of animals are often attacked with disease, while the other parts remain healthy; and light skinned animals are most troubled by flies and parasites." (*Popular Science Monthly*, November, 1885.) The above is worth careful study by breeders. But black has not always been the general color, even among the Aberdeen-Angus. We find, besides black, brown, red or yellow, and dun were the normal colors of the breed at the beginning of the century. Dr. Skene Keith, 1811, says: "the colors which are considered as good are *brown, black, brindle or dun*, if not too white." *Dun*, it is interesting to note, was the color of the fairy cattle (which it would seem also were polled), of the mediæval, superstitious Scot. Mr. Headrick, in the Survey of Angus, 1813, gives the colors as "dark brown, or black, or brown brindled with black. A few white spots, as they give the animal a showy appearance, are not objected to. But if a great proportion of the animal be white, and if, in place of brown or black spots, she be dotted or variegated with blue, she is universally disesteemed. They also prefer these animals with a shaggy but soft pile, as they are best adapted to endure the rigors of winter. But the pile of these animals increases with the cold to which they are exposed." Speaking of *pile*, it should be noted that there are two coats in the Aberdeen-Angus—an under, soft, close, mossy, seally layer, and an upper and longer, silky, thick and flowing fur. When full grown, in winter, this stands up between the spread fingers; the pile is so rich and thick the form is almost completely obscured and the animal might be mistaken, with such an overcoat, "for a bear." The hair should be of the softest, long and silky, but not hard, wiry, stiff or curly. As the hide enters into the composition of the pile, we observe that this item, in the official tests of the Chicago fat stock show, weighs *least* of any breed, and it is therefore curious to observe that in the Aberdeen market the "polled" hides bring 1d. per lb. more than the "horned."

CONNECTION WITH ANCIENT URUS ACCOUNTING FOR VARIATION.

It is a well-established fact, the proof of which we need not here go into, that the wild white Urus, which has still its descendants in the Hamilton Park polled cattle,* roamed over the north-eastern peninsula of Scotland. Indeed, the present Aberdeens inherited many characteristics from the former. Among such inheritances is the variation in color. A little white has never been disliked if properly placed on the under line.

THE YELLOW—" SILVER-YELLOW."

This color, as already seen, was a thoroughly pure color. In fact, Bowie and Fullerton used to declare it was the purest. A bad red is never seen, and a red

---

*" Cadzo Castle, Lanarkshire, the seat of the Duke of Hamilton, with its park, originally formed part of the great Caledonian forest, where King Robert Bruce, according to tradition, hunted the wild bull, in 1320, and where, two centuries later, James IV., of Scotland, indulged in the same wild sport."— J. E. Harting, F.L.S., F.Z.S.—" British Extinct Animals," p. 299.

never breeds a bad one. Among some classes of breeders on this side there would be a decided preference given to "Red Polled Angus"—Hon. M. H. Cochrane, indeed, has been endeavoring to establish a herd of this kind. He has now eleven head of females, and has just bred a red bull of high blood to eventually top the herd. He exhibited some of them in Canada, last fall, and they were much admired. Others have also been experimenting in the same direction. But *black and bonnie*—sleek, sealy, smooth and shiny—is the best looking:

> *All that you wish that's good and comely,*
> *Shines forth supreme in black and humly.*

"THE BLACK POLLS WILL FILL THE LAND LIKE THE BLACK HOGS."

The following remarkable observations from the *Rural Almanac* (*The Field* office, London, 1882), on the subject of "black," are of such interest as to be worthy of place here.

The wonderful prices made in the north-east of the island for the Aberdeen Polls, and in the west for Horned Welsh cattle; in the south and the midlands for Blackfaced sheep and Berkshire pigs, indicate a general leaning to the belief that blackest skins must mean primest meat. Now, it is a fact—a noteworthy one—that *white points do coincide with greater size and coarser fibres*, and that the dark gray or black points do accompany shorter fibre in flesh and compacter forms. How far this tendency can be traced we do not know; it is a subject which requires a Darwin to deal with it. But that there is connection between color and condition of carcass—and perhaps with flavor—seems quite certain. And public opinion —that instinctive, fitful, blind groping in search of absolute truth, which is continually escaping from the grasp—just at present is compelling both Great and Greater Britain, in making purchases, to prefer black live stock. It should be said that this inclination to play rouge-et-noir with cattle, and to put the money down on the black, is not wholly without a justification. The black breeds have made a real advance in recent years; and, perhaps, a greater proportional advance than breeds of any color.

BLACK THE RESULT OF FANCY.

In the case of Aberdeen breeders, they began to prefer the "niggers," thinking they were the hardiest and best beefers. This grew into a "craze," and thence into a fashion, and that settled the matter ever after. The Aberdeen Polled men determined to stick to one of their normal colors, the same as if the Shorthorn men had adhered to an unbroken red, a constant roan, or a whole white. In either case one or the other would, undoubtedly, less or more frequently appear.

"Black" cattle has been the common term to describe the bovine species generally in Britain. But some have suspected that "black" does not refer to color at all, but is a modification of the word *block* cattle, cattle for the butcher, in contradistinction to "cattle" for draught—horses, these being in old times also called "cattle."

Black cattle, however, were, from the earliest times, most generally esteemed, along with the dark reds and browns. Varro, and others, preferred them; and we find our own early writers did so likewise. "Compilers of cattle works repeat, one after another, that the best English oxen and cows * * * are *generally black*" (Lawrence, On Cattle, 1805). Markham (who flourished in good Queen Bess's time) said of these black cattle, "they whose blackness is purest, and their hair like velvet, are best esteemed." This breed, says

Lawrence, " is probably the same as in some parts of Scotland," in his day.

SAMPLES.

The following description from the " Herd Notes " of the *Banffshire Journal*, for January 19, 1886, may be read with interest as being curiously similar to some of the foregoing ideals :

> There has just been presented to the Banff Museum (made famous by Thomas Edwards, F.L.S., and Dr. Smiles' book), the head of a very beautiful young polled cow. This fine cow, which was bred and owned by Mr. Hannay, Gavenwood, was one of the family of Pride of Aberdeen, being Pride of Aberdeen 26th (4560); but her pet name in the herd with her breeder was Lady Paramount, a name which was very well merited. This fine young cow died in February of last year. Lady Paramount was characterised by exceeding sweetness of disposition, beauty of countenance and neck, with the true Pride ear superbly set, immense substance on the finest of bone, the touch being velvety, hair long and abundant, and skin like a lady's glove.

The *Illustrated Journal of Agriculture*, of Montreal, gave an account of the Rougemont Polled cattle, which said the marvel of the herd was Judge, the polled bull bred at Ballindalloch :

> He is the same that won the first prize of his class at the Paris Exhibition of 1878, where the same breed won the championship of the world. Judge is a remarkably fine specimen of his tribe. His measurement is as follows : Girth behind the shoulder, 79 inches ; length from point of shoulder to setting on of tail, 66 inches. His length is prodigious. There is no waste about him, and the thickness of his rounds of beef, his masculine head, his rich coat, level crops, and his wonderful hide have no more bone to support them than is absolutely necessary. His touch is like that of a very well bred Shorthorn.

This remarkable bull died in the possession of Judge

J. S. Goodwin, of Beloit, Kansas, who is quite a patron of the Polls. Mr. Goodwin had the head preserved.

The Smithfield, 1879, "reserve" for the championship, a Scotch polled heifer, was, I observe, described thus: "Her shoulders were covered to perfection, giving her the appearance of having no neck."

An account of Mr. Stephenson's heifer, of 1884, thus hits her off: "Were the hind legs to be cut off at the hocks, the forelegs just above the knees, and the head at the throat latch, the entire animal might be packed into and would fill a rectangular box of proper dimensions."

COMPARISON WITH THE FIRST BOVINE STANDARD, WRITTEN 600 B. C.

The above descriptions may be, with interest, compared with that given by the first to compose a scale of bovine points—Mago, the Carthaginian, who may be correctly termed the father of bucolic literature. Mago lived 600 B. C., and his writings were much admired by all the early Latin writers. Varro, Columella, and Palladius copied his description as the standard authority: "The oxen that we should procure should be young, square formed, with large limbs, high, strong, black horns, forehead broad and curly, ears rough, eyes and lips black, nostrils turned up and wide, neck long and muscular, dewlap large, reaching nearly to the knees, chest broad, shoulders large, belly roomy, and as it were filling out (*barrel shaped*), flanks extended, loins broad, back straight and even, or slightly depressed, haunches (*buttocks*) round, legs compact and straight, but rather short than long, knees moderate, hoofs large, tail very long and hairy, the hair of the whole body thick and short, the color red or dark brown, and the whole body very soft to the touch or handle." "Certainly a very tidy ox, whether he be purchased in Lybia, in the year 600 B.C., or in Northamptonshire, A.D. 1850." It is a description that for twenty-five hundred years has been copied with less or more variation, representing many grotesque burlesques on the typical bovine form. And down to a century ago, and everywhere, except in Britain, where necessity developed a new type—a more trim pattern—it would satisfy the idea of naturalism, which is the embodiment of Prof. G. T. Brown's *muscular maturity* advocacy.*

---

*"Life on the Farm.—Animal Life"; by Prof. G. T. Brown (Chief of the Agricultural Department of the Privy Council): Bradbury, Agnew & Co., Bouverie St., Strand, London.

CHAPTER III.

The Milky Way.

DAIRY PROPERTIES.

The first thing that brought the Aberdeen Polled into notoriety was the famous dairy properties of the dams. The Buchan cattle were sought out on account of this fame. That had spread all over the North, where they were wanted "for the purpose of the dairy" by such men as the late Sir John Sinclair, Bart., chairman of the Board of Agriculture, and projector of the Statistical and Agricultural Surveys of the Counties, the Duke of Richmond, and others. Aberdeenshire was then a great exporter of butter, and it commanded the highest price per lb. in the Edinburgh markets. The export was chiefly from Buchan. The Buchan cows were equal, according to Youatt, to the Ayrshires, as milkers, a sufficient proof of their claims on this head; while the richness of the milk was far superior.

SOME RECORDS.

The following record, for season 1845, is taken from the *Scottish Farmer*, May, 1846. It relates to the cows of such a show herd as that of Robert Walker, Portlethen, and will give some idea of the milking qualities of this breed of beef cattle:

The following account of the yield of milk last season in twelve dairy cows has been furnished to us by a well-known breeder of this race of cattle, Mr. Robert Walker, Portlethen. The yield is given in Scotch pints; a Scotch pint, we may explain to our English readers, is equal to about three imperial pints:

	Pints.	Remarks.
BROWNMOUTH, 7 years old	3024	has had twin calves three times and been but once dry since she calved first, October, 1839.
AUCHLUNIES, 8 years old	2931	gained the second prize as a dairy cow at Aberdeen in 1845.
LADY, 6 years old	2388	pure Angus breed, by Mr. Mustard.
MUSTARD, 6 years old	2388	a good breeder.
COWIE, 7 years old	2388	has twice had twin calves.
YOUNG DUCHESS, 3 years old	1741	first calf.
YOUNG COLLBONIE, 3 years old	1561	first calf.
YOUNG PITYOT, 4 years old	1500	second calf.
EPPIE, 7 years old	2020	has had seven calves.
DUCHESS, 11 years old	1830	the disputed cow at Aberdeen, 1844.
QUEEN O' MAY, 4 years old	1800	second calf.
YOUNG BROWNMOUTH, 3 yrs old	1530	first calf.

The above cows are all black or brown polled, of the Aberdeen or Angus breed, and four of them gave milk until they calved this year, 1846. Their milk is of the best quality, and in general they are good butter cows. The above twelve are the best milkers selected from seventeen.

Of recent breeders, the only one who, owning a crack show herd, determining at the same time not to miss the dairy qualities, was the late Lord Airlie, K. T. Whenever he made a private purchase of a cow, he was particular that she was a deep milker. This is the result, as given by his lordship to the *North British Agriculturist*, December 26, 1879: " I have at present seventeen pure polled Angus milch cows in my dairy The greater number of these give twelve to fourteen, and sometimes sixteen, Scotch pints for a considerable time after calving. The milk is admitted to be much richer than that of either the Shorthorn or the Ayrshire. As regards the length of time for which they will continue to give milk, my cow, Belle of Airlie, (1959), dam of Belus (749), used to be milked all the year round. Last year, when I was away from home, they left off milking her about a month before she calved, and she died of milk fever, induced, as I believe, by the circumstance that she had not been relieved of her superabundant milk. The cow, Miss MacPherson, of the Erica tribe, is now giving six Scotch pints a day, more than nine and a half months after calving." Later he reported having some cows giving as much as twelve Scots pints, or eighteen English pints, daily, though over three months calved.

Adverting to this subject, an authoritative correspondent to an American journal, in describing some sales of herds of Aberdeen Polled cattle, which breed had been "almost exclusively bred for the fat stock market," wrote, June, 1881: " But it is pleasing to be able to point to at least one herd of polled cattle, as an instance of how possible it is to cultivate successfully the dairy properties of Aberdeen-Angus cattle. The herd referred to is one which was founded, in 1848, by the late Colonel Gordon, of Fyvie, Aberdeenshire, and which has been carried on very successfully since that to the present time. The main aim that was kept in view in building up the herd, was to obtain, through careful selection, a race of animals that were likely to prove useful in the dairy; and as the cattle were not forced for show-yard purposes, this ideal was realized more successfully, perhaps, than in any other polled herd in this country. * * * Without exception the cows were heavy milkers, having beautiful udders aud great milk veins."

BUCHAN PRIME IN BUCHAN RHYME.

The following, in the broad Buchan dialect, may amuse some American Caledonians. The lines are by Mr. Colin Macpherson, of Dundee, Forfarshire:

> I'd ance a Buchan coo,
> As black's a craw, an' better too,
> Than ony Ayrshire that e'er cam'
> Frae oot the Wast or Buckingham,
> Or ony shire, I'm safe to sware,
> Her like I never will see mair;
> Her hide was saft as velvet silk,
> An' fourteen pints o' guid thick milk
> She ga'ed me ilka day for lang,
> An' what was better, by my saug,
> The hale yeer roun'—she ne'er gaed yeal!
> Ye need na lauch, my crusty chiel;
> As true as ever hoves were halv'd
> She milkit till the week she calv'd.
> They're mair than me can tell the same,
> The Buchan kye were kye o' fame.
> My neebor man, auld Geordie Garrow,
> Had ane for twal years ne'er fell farrow,
> An' wi' her baith did ploo an harrow.
> She wrocht her wark and milkit weel—
> Some foppish farmers ca'd him feil;
> But, saul, he had mair sense than them
> An' kent the worth o' his black gem.
> He let them see that his coo Keat,
> Baith wrocht an' milkit for her meat.
> Oor guid black Buchan kye, ah, man!

> Were breeders rare an' milkers gran'.
> I've seen a new fa'n Buchan's calf
> Far bigger than ane an' a-half
> O' ony Ayrshire e'er ye saw—
> The Buchan kye nane can misca.

RICHNESS OF THE MILK.

The annual value of the dairy produce from Buchan, in 1810, was £20,000 ($100,000). The richness of the milk is very great. Baron de Fontenay, an observer of the breed for several years, declared that the milk was equal, if not superior, to that of the Brittany cow—and she is the producer of the richest milk in the world.

GENERAL CONSIDERATIONS.

These milking qualities were, however, to be obliterated by the other even rarer qualities of extra beefing that the breed possessed, and that were to become the more favored by the breeders. The deep milking quality was the foundation, however, of the improvement towards the beefy supremacy. It still is the foundation of its proclivity to precocious maturity. The property is, therefore, strongly inherent, and if the breeders want to have herds of dairy polls the attainment of their desires would be an easy matter—as was found by the determined Lord Airlie. Strains of milkers could easily be established, that would not impair the beefing qualities of the produce. But the property is not generally cultivated; in fact, it is not now generally wanted. It would not pay owners of high-class herds to starve their calves for the sake of a few pints of extra milk—even if they had the best ways of doing with it. Treated as beef cattle, the western ranchman has a breed that is quite satisfactory in all that regards lacteal development, in that the dams have plenty of milk to feed their calves successfully, and are easily dried.

The following are three quotations from an article that appeared lately in the *Daily Free Press*, Aberdeen, Scotland, coincidentally with an article on the same subject in an American journal, and may be taken as confirmatory to the above. The *remarks* relate to the native Polled cattle of the county:

(1). Though our cattle breeders are apt to imagine that their pursuit of improving stock is quite a modern thing, *it was pursued, and pursued with intelligence, before the middle of last century*.

(2). While they had regard to "points" bearing on fattening, they did not overlook those that bore on milking qualities.

(3). The other evening it was my privilege to have under my roof a gentleman formerly well known as a breeder of Shorthorns in the north, though now farming independently elsewhere. And amongst other points that came under discussion, I had the satisfaction to find that he held as strongly as myself, that Aberdeenshire breeders, as a rule, had erred greatly in neglecting the milking qualities of improved stock in working up to their beef ideal. This gentleman's opinion is, on every ground, of far greater weight than mine.

CHAPTER IV.

Vigor.

BREEDING QUALITIES.

The breed is very prolific. The cows are roomy; they are milky and good fosterers. Twins and twin producing families are not uncommon. Triplets are by no means rare. Mr. J. G. Walker, Portlethen, exhibited at the centenary show at Edinburgh, 1884, of the Highland Society, triplet heifers, named Asia, Africa, and Australia, which created much interest. Dr. Fleming, the head of the British Veterinary Department, in his great work, " Veterinary Obstetrics," quotes the unique record of a Buchan polled cow on the farm of Balfluig, Alford, which produced twenty-five calves in six years, of which she produced seven in one year. Though seven of the twenty-five did not come to maturity. an average of three calves per year did.* Cases of heifer calves being accidentally got

---

*Quoted from " Veterinary Manual"; by James McGillivray, M.R.C.V.S., L. & E., Rayne : Aberdeen, 1861.—p. 280 : " I shall record what I believe to be an unique case of a calf-producing cow. ' Memorandum regarding a small cow of the black polled breed, which belonged to the late Mr. Alexander Stephen, Farmton, Alford :

in calf occasionally occur, and they have no difficulty in calving, as yearlings. One year a breeder had twenty-seven *queys* served, three being yearlings. Only two broke service, which were easily made fat; the other twenty-one all produced calves the following year. Bulls have been in active service up till thirteen and eighteen years old. Hugh Watson's Old Grannie (1) lived to be thirty-six years old—the greatest age to which a bovine ever attained—and produced twenty-five calves; " Black Meg" was slaughtered by the butcher at twenty years old, and was "good beef." Charlotte (203) was slaughtered when eighteen years old. One of Mr. Ferguson's Princess cows was exhibited at the Highland Society Show when twenty-one years old. Ruth (1169) went to the butcher at eighteen, having been a splendid breeder; and Mr. Anderson lately wrote that "the dealer who purchased her declared she did not look more than twelve," which statement was corroborated by Mr. Robert Bruce, Great Smeaton, Northallerton.

CONSTITUTION.

These facts prove the vigor of their constitution.

Year.	Number of Calves at a birth.	
1842	1—first calf.	
1843	3—came to maturity;	⎫ in one year.
1843	4—one died;	⎭
1844	2—came to maturity.	
1845	3—came to maturity.	
1846	6—died prematurely.	
1847	2—came to maturity.	
1848	4.	

"'One of the above queys was sold for breeding, and produced twins at first calving. The cow was sold at Mr. Stephen's roup (auction), to General Byres, of Tonley, Tough. She had one calf when in his possession.'"

Mr. T. W. Harvey, of Turlington, Nebraska, writes: "The constitution of the Aberdeen-Angus is hardy. They seem to retain the ruggedness of their Highland progenitors, and to have been bred to develop the best feeding qualities. In our importation, we found, after bringing them West, that they did not lose, as most animals do, and have to become acclimated, but from the day of their arrival seemed to thrive and grow. We noted, also, the length of time they could be kept on one pasture, and that, perhaps, a pasture that had been exhausted by other cattle. They seemed to have an extraordinary nutritive system, requiring comparatively little food, and appropriating and thriving on it, with very little waste. They seem subject to few sicknesses, and I have not observed them to show any sensitiveness to the extremes of either the heat or cold of our climate."

SUPERIOR TO THE HARDENING CONDITIONS UNDER WHICH THEY HAVE BEEN REARED.

That well informed journal, the *Farmer's Review*, October 13, 1884, said: "It can be claimed for the Aberdeen-Angus, and it has been proved, that they are second to none for early maturity and hardiness of constitution. * * * In hardiness of constitution they are the acknowledged breed in the uplands of Banff and Aberdeen, where the Shorthorns could not live. * * * There is a wonderful difference between the cold blasts of the north-eastern counties of Scotland and the moist climate of the west coasts—man or beast that can stand the north-east of Scotland blasts can winter anywhere."

It has been asserted that the cattle are closely

housed during winter in Aberdeenshire. Now, this is a half truth. Where that has been done, it has only been for economy. It must argue a specially vigorous constitution that could, after such a winter's housing, as is alleged, stand the raw, biting, searching blasts of the north-east—such blasts as are *never* felt in the *mild*, moist south-west, in spring, early summer, and early fall. The old-fashioned "byres" were, however, more airy and draughty than close and confined. But it is a fact, pretty well known, that the late Mr. McCombie, and all of his time, did *not* prefer to house their finely bred cattle in "close" byres or stables. Mr. McCombie, indeed, was the champion of exposed *open courts* for stock *all winter*. In season and out of season he advocated these *open sheds* against the modern idea of covered courts. Any one visiting Tillyfour in winter, in the height of a storm, could not but have been impressed with the unconcern of the black polled occupants of these celebrated folds, busy over their frosty, snowy turnips, in long rows, close packed, the storm having free vent over them. The turnips taken internally, in such state as they often of necessity had to be in the winter time in the northeast, must have had an even worse effect than exposure externally to the wintry blasts. These sheds had no doors, but were open all in front, and were not on the sunny side, but away from and cut off from the sun. In *Cattle and Cattle Breeders*, Mr. McCombie mentions that he always preferred cattle from the open straw-yards, as they, when turned out to grass, at once "took a start," and never lost it.

So it is a total mistake, whatever "modern improvement" in construction is suggesting in Scotland to-day, to suppose that the breed was thus tenderly coddled and handled with care during winter. The object of the great breeders tended all the other way; and they always protested against anything else in the treatment of their breeding stock.

Notwithstanding this early hardening treatment of the breed, this natural severity did not tend to "coarsen" the hide or hair; it made the latter only more protective. The superior quality of the breed prevented any coarseness. Depend on it, increased coarseness of hide and hair, resulting from *equable moist* climate, as in the south-west, is developed at the expense of the beefing qualities—and a breed that is only capable of thus producing increased coarse hide and hair, under such climatic conditions, should be avoided as a squanderer of nature's gifts.

The winter season in the north-east is not the coldest or severest. It is the *raw, harsh spring*, when cattle *have* to go forth and forage.

The Aberdeen-Angus, of any improved, practicable breed, has been from time immemorial reared in an atmosphere full of the "roughing it" object, yet they are still superior to it, and unapproached in early maturity!

PRINCIPAL WALLEY AND MR. C. STEPHENSON ON THE DISEASE RESISTING CHARACTER OF THE BREED.

Prof. Walley, P.R.C.V.S., Principal of the Royal (Dick's) Veterinary College, Edinburgh, states in his work, "The Four Bovine Scourges," that "the Aberdeen-Angus are totally free from tubercular troubles that have caused such wholesale ruin in other cattle"; and Clement Stephenson, F.R.C.V.S., Chief Veterinary Inspector for Northumberland, has recorded the same:

The first season I had pedigree polled cows I was much struck with their aptitude to fatten. They were grazing in the same fields with other well-bred horned cows (all were suckling calves) and, while the blacks were full of flesh and in splendid condition, their fellows were so lean that I had to instruct my bailiff to give them a liberal supply of cake.

The more I see of this breed of cattle, the more I am convinced of their great value; they are, it is well known, able to live and look well on a poorer class of land than many other breeds, and yet they repay, in a very marked degree, any attention they may receive, either by putting them on good land or giving them extra feeding.

There is another and most valuable advantage these cattle possess, namely, their remarkable freedom from tubercular disease, a disease that has caused great loss and made sad havoc in many a herd, and a disease the importance of which, in a medical point of view (viz., its communicability to man), is now attracting much attention. Of course, I cannot assert that it has never been known or seen in this breed of cattle; but this I can say, that although I have had special opportunities for research, and have examined great numbers of cattle, both alive and *post-mortem*, I have never yet seen a trace of it in this breed.

In *The Live Stock Journal**—which every stockowner ought to read—for February 12, 1886, is an article, " Muscle: Constitution: Fat," which we consider of the highest importance, especially in connection with the questions agitating the minds of British breeders just now. These questions have become accentuated in the cry " Are British cattle degenerating?" and have arisen from the particular prominence given to the dangers of this very dread disease —tuberculosis—by Prof. G. T. Brown, in his book already noted. We have therefore no hesitation in giving the essential portion of the article. It conveys a message of pregnant significance, apparently from a writer of the highest possible authority:

.* Published by Vinton & Co. (Ld.) : 9 New Bridge St., Ludgate Circus, London, E. C.

To return, what lessons can we learn from these hard times ? We, in the first place, must now be taught that the butcher's stall is the testing-place of all beef cattle; and therefore we must breed animals that, when killed, will look well on the butcher's stall, please the consumer, and in consequence be sold readily. Have we Shorthorn men been producing such an animal ? I fear not. Style and character and such like nonsense have "done away with us." I can remember while suffering from the "fever," agreeing with a Shorthorn judge that a Stratton champion at London was all very good "and that sort of thing," but wanted sadly the style and character of a true Shorthorn. What had we got ourselves to believe ? Simply that a rounded level frame, such as Lady Pamela has, full of lean flesh or muscle, with light offal on small bones, was not what we wanted to breed ! We seemed to believe that the early Shorthorn breeders bred animals suitable for our times. Everything points to the fact that these earlier Shorthorns were "all hills and holes," very suitable, no doubt, to cross with the hard-fleshed, slow growing cattle of the time, but no use to us now, when we have had infusion after infusion of a "tendency to fatten blood" into all our common stock for the past fifty or one hundred years.

We have bred from animals with a "tendency to fatten" till we have in too many cases lost both flesh and milk. Now, I am one of those who believe we can produce Shorthorns that can both milk and lay on good flesh and fat with any breed under the sun ; and I also believe these hard times will bring Shorthorn men to their senses, to the benefit of the breed. What has given the Polled Aberdeen-Angus men such a pull ? Simply this—they have a breed of cattle full of lean flesh or muscle, an animal that, when killed, pleases the consumer, and therefore the butcher. Consider what a record the Aberdeen-Angus have made at the past Christmas shows and sales. I do not refer to the wonderful heifer bred and shown by Mr. Stephenson, so much,[when I speak of the record the Aberdeen-Angus cattle made, as to the fact that Aberdeen-Angus and crosses from them had it almost "all their own way" at all the principal shows, as pointed out in the *Field* a few weeks ago. Take the cross-bred classes in London, for instance—classes judged by the three Shorthorn judges—and then we find blacks and blue-greys heading each class, and in more than one case taking all the prizes in those classes.

There must be some reason for this ; in my opinion it is not far to seek. It is, no doubt, because the Aberdeen-Angus and their crosses have more flesh or muscle than our Shorthorns. We have bred too many of our Shorthorns to death—we have lost that amount of muscle which, in my opinion, is proof against that fearful disease Professor Brown speaks of. I quote his own words: "Tuberculosis, which is a similar disease to consumption in man, extends its area every year among our best cattle, to the risk of the extinction of the variety, and the great damage to public health." And again he says: "Tuberculosis, a disease which is extending year by year in some of the cultivated breeds of cattle, threatens serious results unless the greatest care be taken to avoid using infected animals for stock purposes." I should much like to put this question to Professor Brown: *Did you ever find an animal*—and by this I mean an animal of the bovine kind—*full of flesh or muscle suffering from this disease*?

From one extreme I may be running to another, but I now firmly believe that every one of those animals that have that peculiar soft handle I was taught by my brethren in the Shorthorn world so much to admire, has tuberculosis in one or other of its stages.* Up to the time that an animal is in the last stages of this fell disease, I believe its handling would delight many of our Shorthorn judges. I have seen such awards made, and heard such nonsense spoken by acknowledged judges even at our national shows, as has led me to question if they really knew the difference between flesh and fat. In many cases I feel convinced they did not. People rave over the so-called beautiful handling of prize Shorthorns, when milk-feeding, fat, and tuberculosis have all to do with what they so much admire.

Much nonsense has been said and written about over-feeding at breeding shows, and one animal in particular figured often in *The Live Stock Journal,* last season. Lady Pamela, full of flesh or muscle, was put aside as being over-fed, and her thinner-fleshed, but quite as fat, and certaicly harder-fed companion from Catterick, awarded the prize. A level, full-fleshed animal is spoken of as over-fed, when a thinner-fleshed one, perhaps much fatter, is spoken of as being in "nice breeding condition." I would say, leave the rules of the shows and conditions of entering for competition as they are, but choose judges who know the difference between flesh and fat. With men who know their business, Collings' White Heifer, as we see her in

---

* The author has also similarly expressed himself.

our plates or pictures, would at once be sent to her stall. We do not want them in these days "all hills and holes," but the animals now needed must be smooth made, full of flesh, with heavy rounds and muscular roasting joints. In the show-yards at present, however, we must, forsooth, submit to having our animals passed by "famous Shorthorn breeders," strong family and breed partizans, who breed by paper pedigree, and, perhaps, never produce an animal with sufficient flesh to make it a good one. One feels queer, to say the least of it, to see them touch the animals over—apparently gloating over milk-fat, blubber, hair, and tuberculosis, and award the prizes to fleshless, useless animals. Do the breeders who understand their business, the veterinary faculty, the butchers, or the consumers of beef see the force of their decisions? I venture to say they don't.

CHAPTER V.

Rangers.

THEY ARE "THE CAN'T-HOOK BREED."

Their utility is great in the want of the horns. The horn is an entirely useless appendage to bovines in domestication. It was an acquired weapon, without doubt. The first bovines had none, assuredly, as they did not require any. The Aberdeens being the prime butchers' beasts, this combination will make them receive wider attention. For droving in close herds, for shipping by rail or over sea, the want of the horn is greatly appreciated by experienced handlers. They can be packed much closer; men can move about them so easily, and they do not tear the hide or do, maybe, worse things to their companions. Along the close, tied up lines of cattle, at London markets, none are so comfortable as the close rows of nobby, Polled Aberdeens. Their heads have freedom, while the long horns of the Hereford, Sussex, or Devon, are frightfully in the way.

It is not alone among their own kind that this feature is valuable. The London *World*, *e. g.*, treating of the celebrated "Jim Lowther at Home," supplies us with an interesting item (October 26, 1882): "At the Home Farm (Walton Castle) are some prime black polled Scots, a class of animals far more manageable than either long or short horns. Not only are they easier to pack, but there is no danger of their hurting a promising yearling when they are crowding together." Mr. Geo. Hendrie, President of the Detroit Street Car Company, was led to prefer the Aberdeens for a similar reason—"they can't hook the colts."

Mr. L. A. Hine, of Erie county, Ohio, writes in the *Western Rural:* " During the three years that we have handled these cattle not an animal nor a person has been injured. The cows and heifers are never tied when stabled, and in the pasture and yards they huddle together like so many sheep. They are extremely docile when properly treated."

E. G. Underhill, a young breeder who has started a nice little herd at Norwalk, Ohio, writes: " The reason I came to prefer the Aberdeen-Angus to the other rival breeds, after posting myself on them all, is shortly told—I liked them because they were humly,* fattened easily and early, and *were as docile as lambs.* I would not part with my pollies."

FEEDING AND GRAZING QUALITIES.

All these points go to make up an easy tempered animal, specially fit for stall feeding. They are grand grazers. It is generally admitted that three Aberdeens can be kept for two Shorthorns.

After trying many breeds as grazers, Mr. McCombie

---

*Humly, the Buchan dialect word for *polled;* meaning the same as *doddie.*

found the Aberdeen surpassed all. They were the "rent-payers." They were the only ones that withstood "the cold calendars of May." They are excellent foragers, and they are giving great satisfaction on the plains.

They graze now all over Scotland and its isles, England and Ireland, in fact, everywhere, where improved cattle are wanted.

IN AUSTRALIA.

A correspondent lately sent a description of one of the polled herds, in New Zealand, to the *Banffshire Journal*, January 5, 1886: "I believe that the breed is eminently suited to the soil and pasturage of New Zealand; in fact, while traveling through the country, I have noticed, once or twice, a polled beast amongst a mixed lot, and he has almost always been the best." In Victoria and Queensland, also, there are some fine herds. Mr. J. L. Thompson, well known in the former, and Mr. Gordon, Inspector of Live Stock, in the latter, have described them as having taken successfully to those arid countries, and stamping their characters through a long line of descendants wherever they go.

ON THE AMERICAN PRAIRIES, ETC.

Mr. James Macdonald, whose name is so well known in live stock circles, by his book "Food from the Far West"—which is an excellent guide to the American live stock industry—in 1877 inspected the Victoria Ranch, Kansas, where some Polled Aberdeen bulls had been imported from Scotland: "The polled bulls were full of flesh, and seemed quite at home on the prairies.

* * * The young stock, of course, are not perfect, but the improvement, especially in quality, is very marked. * * * Considering that, even in the fiercest day in winter, they had no shelter and no feed but what they could find on the open prairies, they were really in splendid condition. All were lively and healthy, and the loss by death last winter, severe as it was, was less than five per cent. This season's calves are nearly all dropped, and it is expected that for every one hundred cows ninety-five calves will be raised. The descendants of the polled bulls are easily recognized; they are nearly all black; few have 'scurs' or horns and in general style and quality they are unmistakable polls. They do not stand so high as the shorthorns' crosses, but are thicker, and, as a rule, more fleshy. The prairies of the West seem admirably adapted for the 'glossy blacks.' * * * Last winter Mr. Grant fed an equal number of shorthorn and polled crosses of his own rearing on Indian corn, millet and hay; and on their being slaughtered, at Kansas City, the blacks were found to weigh, on an average, 100 lbs. per head more than the roans."

In 1870 Mr. McGibbon, chamberlain to the Duke of Argyll, purchased for the West Indian Company of that pestilential island, Demerara, six bulls of different breeds. What was their fate? All died but the Angus bull, and when Mr. McGibbon left the island, two years ago, he was vigorous, multiplying and replenishing the earth.

The same may be said of the bulls sent to Jamaica, Buenos Ayres, etc., by the Hon. Charles Carnegie, from the Southesk herd

In the *New Mexico Stock Grower*, of February, 1885, we read what Gov. O. A. Hoadley says of the Aberdeen-Angus:

They face the storm and graze unconcernedly, while other range stock are humped up and cold. They seem to enjoy the snow and root it like hogs. I have fed none of my polled stock this winter.

W. T. Holt, of Denver, Colorado, observes of them, comparing them on the range with his many thousands of different breeds :

The Aberdeen-Angus will outfeed all other breeds.

Mr. W. H. Embry writes the Kansas City *Live-Stock Indicator*, as follows :

As a matter of interest to Western cattlemen, I will give you my experience in feeding cattle last winter. I fed nineteen half-blood Angus, eight half-blood Herefords, and twenty-five high-grade Shorthorn two-year-old bulls at the same troughs. The Angus bulls stood the storms better, took on fat faster, and were in better condition in the spring, than any of the other breeds. I found the absence of horns a great comfort to the animals, as well as satisfaction to the feeder, not simply on account of taking less room at the feed trough, but because it saves the animals much worry and many bruises. While the muleys were able and ready to defend themselves when attacked by the horned bulls, they were more quietly disposed. I have not been prejudiced against nor partial to any breed of cattle, and make this statement simply from my impartial judgment and experience.

Judge J. S. Goodwin, of Beloit, Kansas, writes to the *Breeder's Gazette :*

General Campbell, of Denver, breeder of Herefords and Angus, brought in an imported Angus bull off the range, in May, while I was there, and told me the Angus came through the winter the fattest of all his cattle. Mr. Alstot, of Lawrence, Kansas, told me last year he bought 380 head of Angus, Galloway, Hereford and Shorthorn bulls for one range in New Mexico, and that the Angus beat them all, coming out of the winter sleek and fat ; the Galloways were second, Herefords third, and Shorthorns a poor fourth.

THE BREED THAT BEATS THE RECORD. 43

The Rocky Mountain *Husbandman* says of the Aberdeen-Angus cattle in Montana:

Until recently but little was known in this country of the Polled-Angus breed of cattle, and the feeling tended largely toward a preference for the breeding of Shorthorns for the improvement of our range herds. But the recent introduction and experiments with the Polled Angus bulls develop many points of excellence for that breed over any other that has ever been tried in our Montana climate. The Montana Cattle Company, we believe, was the first to turn pure bred Polled Angus bulls upon our range, and we learn that the result has been highly satisfactory. The breed seems to be stronger than our native cattle. It is noticed that the calves from our native cows, bred to Polled Angus bulls, invariably take after the sire in color and form, so far as that they have no horns. This is one of the most superior points in the Polled Angus. Cattle raisers who have devoted much time with our range herds now concede this is a great advantage over horned cattle. A considerable loss from death, yearly, comes from the freezing of horns among our common herds. Frozen horns do not cause instant death, but affect the condition of cattle severely in the spring, preventing many from taking on flesh readily, and often causing such mental derangement that tame cattle become unmanageable. Those who have been engaged upon our ranges for some years, and have given close attention to our herds, estimate a yearly loss of one per cent. from this cause. Cows heavy with calf are more easily affected than other cattle.

Messrs. Martin & Myers, extensive stockmen of Shields River, fully concur with us in the superiority of this hornless breed. A year ago they brought out sixteen Polled Angus bulls, thirteen cows, and four yearling bulls, which they turned out among their cattle. No attention was given them through the winter, and from the account we have from the foreman of this herd, the cattle did not need any. Of the thirty-three head all came through the winter in fine condition except one, which was lost by accident. It was noticed that in the severest weather, when other cattle were seeking shelter from the cold winds, these cattle would go upon the highest and most exposed ridges to graze, as perfectly unconcerned as if in a warm climate. One young bull, which was very poor when turned out, wandered away, and was not seen during the winter, and it was thought that he had probably got snow-bound and perished; but, in

April, to the surprise of all, he came down from the hills in first-rate condition, showing that the bracing atmosphere of our Montana climate had agreed with him, and that his condition was really better than when he went away, six months previous. The result with the Polled Angus shows a great superiority over our common stock cattle, and more especially Shorthorns and grades which were brought from the same State last year and wintered here, the loss among which was very great.

The *North-Western Live Stock Journal*, published at Cheyenne, Wyoming, also notes these cattle:

> The Messrs. Myers, who own a large ranch with several thousand head of cattle ranging on the Yellowstone river, some years ago put into their herd a number of well-bred Polled Angus bulls, at the same time placing in the herd a number of Polled Angus cows. They have now a very large number of black polled calves, and so well pleased are they with the young stock that they gave an order for the purchase of the animals they have just received from Messrs. Estill & Elliott. Messrs. Myers, believing in the superiority of the Polled Angus blood, propose to give their entire herd a liberal infusion of the best strains, and as rapidly as possible produce a herd of black polls. They take this course because, they say, "they believe, first, that the Polled Angus is the best beef animal, and second, that he is the best rustler."

The following is from an article in the London *Field*, entitled "Cattle Ranching in the Great Lone Land":

> On the Cochrane ranch, however, they are now using a number of thoroughbred bulls—Shorthorns, Herefords and Polled Angus—preference being given to the last as the hardiest and best suited to the climate. These bulls were brought from the Cochrane herd in Lower Canada; they were conveyed by rail and boat to Fort Benton, on the Missouri, 3,000 miles to the ranch, and then driven overland 400 miles. This journey afforded a complete test of the stamina of the respective breeds. When they arrived on the ranch, on Bow river, the Shorthorns were barely alive, the Herefords had suffered less, but still were greatly reduced. The Polled Angus reached their destination in good condition, and during the severe winter of 1882 and

1883 again proved the superiority of their constitution, and their perfect adaptation to their new home.

This article is signed by a Mr. J. A. McMullen, whose name is unknown in Aberdeen-Angus circles; but at all events the accuracy of his statements can easily be verified or confuted by readers themselves.

Mr. Geo. Findlay—of Messrs. Anderson & Findlay, Lake Forest, near Chicago, Ill., who own, perhaps, the oldest herd of Aberdeen-Angus cattle in America—writes to us: "I have had many conversations with ranchmen, who have tried them in Manitoba, Wyoming, Colorado and Texas. From these, I am convinced that they are unequaled for this property. As we have just gotten through with a spell of very severe weather, I am reminded of the claim made for them by a Montana ranchman, whe has tried them. It was this—that the horn is very sensitive to low temperatures, and very readily freezes, so that, in a horned herd, there are very frequently found, next season following a severe winter, a great many animals with one or both horns drooping—the result of being frozen the previous winter. These animals are invariably in very lean condition, and in a precarious state of health. You can take this statement for what it is worth. Another Montana ranchman told me that his Aberdeen-Angus, of all the other breeds he had, were always the first in the morning to leave the shelter and forage for grass, no matter how stormy or cold."

Mr. J. J. Hill, of St. Paul, who owns large and fine herds of Aberdeen-Angus and Shorthorn cattle, writes: "If inquiries for Polls is any indication of their doing well in the North-West, they surely are the cattle for this country."

The *National Live Stock Journal*, May, 1883, says:

"Of the *Black Polled* Scotch cattle * * * the Angus, or Aberdeen, have taken precedence. * * * The bulls answer admirably for crossing with females of other breeds, and especially those of Texas, as their calves come of a black color, and grow up hornless, or with mere stubs or scurs. Thus far they have proved themselves very prepotent."

The same number contains an item which says: "They endure the sudden changes of weather, and especially the severity of cold, and come out in better condition than any other imported cattle."

S. P. Cunningham, in the July number, also refers to the necessary use of the Polled Angus on the ranges. Again, November, they "are particularly sought after for breeding on the vast plains bordering the Rocky Mountains."

The *American Agriculturist* produced in its February number of this year, two remarkable and finely executed engravings, which the following extract, from the article accompanying them, will graphically explain: "Our artist, Mr. Forbes, gives a picture (page 51) from real life of the introduction of an Angus bull upon the plains. The stalwart herders eye him critically from under their sombreros, taking in his fine proportions, and perhaps seeing in the dim future essential changes which he is likely to introduce in their wild modes of life. Wild or half-wild cattle, need wild or half-wild herdsmen. The polled grades are quite docile, easily herded, neither liable to injure themselves, the cowboys, nor the horses; they grow rapidly, fatten kindly, and fairly load themselves with flesh when pushed. This Mr. Forbes shows admirably in the group of ripe-fat steers in the lower picture, ready for the poll-axe. Two have mere apologies for horns, loosely attached;

most of them are polled. They can be shipped from Abilene, Kansas, or Denver, if necessary, to New York, taken out of the cars, fed, watered, and re-shipped, every few hundred miles, and arrive showing very little shrinkage, no fever and no wounds; and besides, they can be packed closer than horned beasts, greatly saving in freight."

The *American Agriculturist* has been consistent in its advocacy of Polled Angus cattle for the plains.

CHAPTER VI.

Grades.

CROSSING AND CROSSES.

The following appeared in the *Agricultural Gazette*, August 8, 1881, on this subject: "A great deal is written and spoken about 'Aberdeenshire crosses.' Shorthorn men seem to find in this a never failing theme to advance the praises of their own favorites. So they well may, for in how many reports of English Christmas exhibitions do we read 'the Polled Scots (Aberdeens) and Aberdeenshire crosses continue to be the backbone of the exhibition.' Certainly, in recent years, these cattle have asserted their position most unscrupulously, and to the Shorthorn do Shorthorn men claim all the merit. This must be objected to. Taking the Aberdeen, the Devon. and the Hereford— we put to a female each, of these, the Shorthorn male. Now, if it was to the superiority of the Shorthorn entirely that any good result was owing, would not the three crosses have been equally as good? But the result is that the Aberdeen cross is by far and away the best. This test proves that it is not to the Shorthorn that the Aberdeen cross owes its excellence, but

STEPHENSON'S CROSS-CHAMPION OF 1884 (See p. 53)

to its own Aberdeenshire dam; and also proves the Aberdeen superior to the Devon and Hereford pure. Besides, if we go far enough back in the history of the majority of the Shorthorn sires used for crossing in the north, even they are founded on a cross with the native." So that some of the fame of the Aberdeenshire Shorthorns belongs also to the Aberdeen. For years there have been classes for "crosses" at all the fat stock shows of England—for years the Shorthorn has been bred in every district in Britain, where there have been native breeds, and these "cross" classes have been open to breeders of Shorthorn crosses from Hereford, Devon, and other breed districts, as well as the Aberdeen district. What is the result?—the Aberdeenshire crosses fill the classes. Mr. G. T. Turner, stock editor of the *Mark Lane Express*, writing as correspondent to the *National Live Stock Journal*, Chicago, says:

> For evenness, thickness, and quality of flesh, together with size, I do not think anything which comes into the London market can excel a first cross between the Shorthorn and Polled Scot; and these, together with similarly bred cattle from Scotch dams which have one or two crosses in their blood, are the best "crosses" which are exhibited at Islington and Birmingham fat stock shows. For example, the awards at Islington, December, 1877, for "cross or mixed bred cattle," were as below:

Steers not exceeding 3 years old.

Cup £40 and 1st prize.. } Shorthorn and a Polled Scotch,* second cross.
2nd " .. Shorthorn and Galloway, first cross.
3rd " .. Shorthorn and Polled Scotch, second cross.

Oxen above 3 years old.

1st prize.. Shorthorn and Aberdeen, first cross.
2nd " .. Shorthorn and Scotch Polled, remote cross.
3rd " .. Shorthorn and Aberdeen, second cross.

> With a very few exceptions, the entries of cross-bred cattle,

---

* "Polled Scotch" means "Aberdeen Polled."

both at Islington and Birmingham, were from Shorthorn sires and polled Scotch dams. I give this detail because it is important that it should be clearly shown and understood that partly-bred, mongrel-bred, or—if I may be allowed to use the American term—"grade" Shorthorns do *not* top the English market. Again : the Shorthorns are never, that I am aware of, quoted at equal rates to the Scots, Herefords, and Devons, in the London market returns ; and the difference in the value per pound is greater than many suppose.

Where were the Devon and Hereford crosses?

THE ABERDEEN DAM IS MORE PREPOTENT THAN THE SHORTHORN SIRE.

The Banffshire (Scotland) *Journal*, commenting upon the appearance of three white Shorthorn bulls in one ring at a late Morayshire show, at Elgin, states that "A fancy is growing in Morayshire for white Shorthorn bulls, to mate with black polled cows. They have been crossed in some herds with the best effects, the produce being beautiful grey or black calves." Adamson says : "The greater part of the crossbreds exhibited at the fat exhibitions, and the crossbred class is the acknowledged feature of the show, is the Aberdeen-Angus cross, and I say, without fear of contradiction, that the money prizes are invariably awarded to that cross." Mr. J. H. Harris, 1871, gained the championship with such an animal—bred by Mr. Robert Bruce, then of Struthers—the dam of which ox was purchased by Sir George Macpherson Grant, with the intention, presumably, of making her the initial foundation of a new family.

In 1880, the champions at Hull, Birmingham, and Smithfield—the three principal fat stock shows of that year—were animals *bred in Aberdeenshire*, sired by

Shorthorn bulls, and out of Polled Aberdeen dams; "a striking tribute to the skill of Aberdeenshire breeders, to the Aberdeenshire cattle, and to the premier cattle producing county of Britain." The Hull champion was bred and exhibited by that well known feeder and breeder, Mr. Jas. Reid, Greystone, Alford; the Birmingham winner was bred by Mr. Wm. Leonard, Farmton, Alford; the Smithfield champion by Mr. Durno, Jackston, Fyvie; and having captured the eye, as a youngster, of Mr. William Middleton, then of Greystone, Alford, was sold by him to Mr. Garret Taylor, Mr. Colman's agent. This ox was as even as a die, and got all his good qualities from the Polled. He was exhibited by J. J. Colman, M.P., the enterprising breeder of Norfolk Polls, and fancier of the Scotch (Aberdeen) Polls. Like McCombie's Black Prince, he was forwarded to Windsor, "by command," for Her Majesty's gracious inspection, before being slaughtered.

A report of the 1880 Smithfield show said: "The cross-bred cattle are conspicuous for vigor, size, and quality; and it is clear that much will be made of them in the future, now that they have entered the charmed circle of Smithfield champions." But this was not the first time that they had done so.

Mr. G. T. Turner, witing of the 1882 Birmingham show says; "The division of the show which I thought by far the best was that of the cross-bred cattle" (January, 1883, number *National Live Stock Journal*.)

Again of the Birmingham show, 1883 (same to the same): "Cross-bred cattle were the best section of the show, to my thinking. * * * The cows and heifers comprised one of the plums of the show which got the cross cup. I liked her next to the champion

Scot," of which animal, exhibited by Mr. C. Stephenson, he wrote: "As a champion prize winner, I have not seen one I have thought more deserving. She was a fine carcass of beef, with the smallest possible bone and least possible inferior meat. I am sorry that this heifer will not be able to 'try conclusions' at Islington, next week. She has won champion honors at Norwich, Leeds, and Birmingham." The regulations debarred Birmingham cattle from going to Islington that year, else, as many competent judges have asserted, the Smithfield, 1883, Champion might have been of the "black and humly" kind produced by Mr. Stephenson.

Mr. Geo. T. Turner, writing of the crosses at this show, said: "The Shorthorn-Aberdeen cross, which has hitherto proved to be the best of all crosses in the show-yard, goes to show greater prepotency on the part of the Aberdeen Polled cow than on the Shorthorn sire, the greater portion of the animals so bred showing scarcely any variation from Aberdeen Black Polls."

Mr. Turner had also written in 1882; "Thus it will be seen that, as far as the cross-bred animals, at Birmingham, are concerned, the Shorthorn sires have not been so prepotent as the Polled Aberdeen dams."

All the above is surely sufficient to uphold the "assertion" first made by the writer in the *Agricultural Gazette*, quoted in the beginning of this chapter, viz., that the "Aberdeenshire cross" owns its excellence *not* to the Shorthorn element, but entirely to the Aberdeenshire dam.

SO NOW THE POLLED BULL IS PREFERRED TO THE SHORTHORN.

Again to quote the article in the *Agricultural*

THE BREED THAT BEATS THE RECORD. 53

Gazette, of August 8, 1881 : "For some time back, however, the use of the Shorthorn sire has given place, in the north, to the use of the Polled Aberdeen sire on Shorthorn and cross cows. This method is giving the greatest satisfaction, and the best crosses we have ever seen have been so manufactured. The result of this is that such cross is at least three-fourths Aberdeen. These facts are highly important to owners of other breeds, who are turning their thoughts to the 'all round' Aberdeen." The above remarks have borne fruit, and have materialized in the outcome of some of the most remarkable classes in show-yard annals, among which has recently appeared, among others, Mr. Clement Stephenson's Birmingham heifer of 1884 —with which he gained the Elkington Cup, which has to be won two successive, or any three years. This heifer was bred as above indicated. She was three-fourths if not wholly Aberdeen, and a perfect butcher's beast. The influence of the polled sire on the Hereford may be seen in the famous heifer—winner last year of the *Breeders' Gazette* Challenge Shield, in American fat stock shows—Burleigh's Pride, an Angus-Hereford. A most remarkable display of crosses of both kinds appeared at the fat stock shows, this year, in England. The champion prizes, besides the prizes in the cross classes, are almost wholly Aberdeen-Shorthorn. It is a most remarkable thing that the Aberdeen-Polled are now almost the only thing that the many best feeders in England will touch, in the attempt to wrest the champion honors at the fat stock shows. A show at Smithfield, Birmingham, etc., is becoming more like an Aberdeen Fair than anything else. Full particulars of the 1885 crosses will be given in the last chapter.

THE POLLED SIRE IS USED EVERYWHERE.

In West Galloway, the writer of the Prize Report, in the last volume of the Highland Society's Transactions, 1885, p. 112, mentions in two places in his essay, the use of Polled Angus "bulls" for crossing purposes —proving that, in that region, the old reminiscence of the breed is still strong.

All over Scotland, the Western Isles, and the Orkneys have herds been started, or bulls introduced for crossing. The same may be said largely of Ireland, and in England they are prime favorites for both purposes, and rapidly increasing. An authoritative correspondent of the *National Live Stock Journal*, of Chicago, thus writes, February, 1886:

I am no partisan of any breed of cattle. I have been bitten by the pedigree craze, and know a good deal about it. It grows on all breeders who study herd-books, and it educates them far beyond "nature's laws." Could you believe it? I have really and truly got myself to believe that we ought to breed such animals, if possible, as Thomas Bates and Richard Booth brought out and exhibited, while round the walls of my little smoking-room, where I pored over pedigrees and settled the exact nicks in breeding, were pictures of animals belonging to those "fathers of Shorthorns" that were anything but the class of cattle I wish to breed.

We must breed for the "block test." All nonsense must be laid aside. Those Polled Aberdeen-Angus men, without doubt, have the animal at present, and they will reap the harvest. I know of one farmer in Yorkshire, the home of the Shorthorns, who, for the last three years, has put an Aberdeen-Angus bull to really good Shorthorn heifers, and five of his first year's crop sold, at ages from twelve to twenty-three months old (two of them to a professional fat-stock show exhibitor), to average him £40 ($200) each, while he refused £30 ($150) each for the balance he has on hand, four animals, from a local butcher. This beats all his neighbours' pure breeding, and for why? Because he is producing a beef animal.

"Hermit," the English correspondent of the Kansas City *Live Stock Record*, writes: "Polled Aberdeens are great favorites in England, and recent sales would show that there is an increasing demand for them in the midlands and in the fen country. The 'black diamonds' will thrive in any country, and it is amazing that English graziers did not take to them long ago. The victories of Mr. Clement Stephenson, at Birmingham and London fat stock shows, has positively directed the attention of feeders to the breed, and they are now disposed to ally themselves with it. Mr. Greenfield, of Dunstable, has sold a bull for service in Cornwall, where it will be used for crossing purposes. The foundation of a herd has been laid by Mr. G. Simpson, of Lutterworth, in Leicestershire, and Mr. G. Townsend, of Fordham, in Cambridgeshire, has also bought a bull for crossing."

Again: "The 'black-skins' are also becoming great favorites in England, and I see another breeder in Warwickshire has purchased a pure-bred Polled Angus bull for service in his herd for crossing purposes."

Almost every paper we get records similar items.

Well might the late H. D. Adamson exclaim: "The polled sire, on Shorthorn and cross cows, has been resorted to with equally good results." He perhaps had in his mind Mr. Stephenson's famous heifer, and what she was capable of doing. These claims have also been fully borne out on American soil.

Mr. Hine writes: "We have crossed them with the Shorthorns with great success, and of about eighty half-bloods that have come under our supervision but two of them failed to inherit the color, while all are hornless."

Mr. T. W. Harvey says: "Writing to a friend of

mine, in Chicago, lately, in answer to a letter asking him for statistics regarding these Scotch cross-bred animals, Mr. Cruikshank says : ' I have no statistics at hand, but we all know that when we want the very best beef to be obtained, we cross the Shorthorn and the Polled Aberdeen-Angus.' In this country, the crossing of Aberdeen-Angus bulls with our native cows has proved remarkably successful. The first Aberdeen-Angus bulls I ever saw in America, were taken to western Kansas and bred to Texas cows. About three years after, I saw a report, in a Kansas City paper, of a shipment, to New York, of several car loads of two-year-old polled steers from these Texas cows, and the average weight, in New York, was given as 1,368 pounds. The evidence of all who have used the Aberdeen-Angus bulls on Shorthorn grade or native cows, is that the produce are nearly all black, and without horns, good feeders, and mature early."

Again, here are two items from the *Canadian Live Stock Journal*, published at Hamilton, Ont.:

Mr. Coates, of Eaton, P. Q., had on exhibition at Sherbrooke two specimens of the first cross of an Aberdeen-Angus bull on ordinary grade cows. Though but a little more than one year old, they weighed between 1,000 and 1,100 lbs., and are fine specimens. Though the dams were roan, they are jet black, without one single white hair, and are fine thrifty beefers. They were sired by Mr. Pope's bull Proud Viscount. Mr. Pope has also many excellent specimens of the Aberdeen Poll grades.

The Geary Bros. Importing Co., Bothwell, had twelve purebred Aberdeen-Angus calves (December 17th), and expect the number to reach sixty head there and at Bli Bro by 1st June next, and of Aberdeen-Angus grade calves about ninety by same date. Of these quite a number will be three-quarter grades. This is a very practical way of demonstrating the merits of this splendid breed.

THE BREED THAT BEATS THE RECORD. 57

We are favored with the following from Geary Bros.: "We bought, in 1883, a number of common Canadian cows, which we bred to pure bred Angus bulls, the result being a very large percentage of calves. The bulls were all sold, but the heifers, which still remain on the place, and are somewhat less than nineteen months old, on the average will weigh nearly 1,000 lbs. each, and that on ordinary keep. One of the above lot, when a calf, got in calf to a pure bull, and calved at twelve and a half months; a fine heifer calf, and now, at the age of nineteen months, she is in calf again, and will calve at twenty-three and a half months —two calves before she is two years old!"

The *Canadian Live Stock Journal,* writing of the show at Sherbrooke, Canada, says:

It was with peculiar satisfaction that we noticed the extensive and painstaking efforts that the Hon. M. H..Cochrane, of Hillhurst, was making in this direction, as witnessed by the many specimens of the cross-breds that were exhibited by him at Sherbrooke, at the Eastern Townships Exhibition held there last October. Some of the various crosses only can we refer to.

Aberdeen-Angus—Shorthorn cross-bred heifer. This specimen, a red in color, two years old, was sired by the Aberdeen-Angus bull Northesk A (1478), and out of the Shorthorn cow Belinda, by Prince Arthur (27506). This heifer had an Angus head, nice sloping shoulders, and was pretty in many ways.

Aberdeen-Angus—Ayrshire cross-bred heifer. Here the Ayrshire color was completely lost in the Aberdeen-Angus black. The sire, Sidney (2360), and the dam the Ayrshire cow Heather Bloom. This one-year was a good beast. The shoulder was thickened, and also the quarter.

Grade—Aberdeen-Angus steer. This portly fellow, calved April, 1884, was a chunk of beef of prime quality, sired by an Angus bull, and from a small native dam. He was red in color, immense on the crops, and good in all his parts, which brought him the gold medal for best beef animal at the show, and first also as best grade yearling steer. We can speak of this cross with unqualified praise, an opinion borne out by the results of the Experimental Farm tests, by the experience of Geary Bros. in the West, that of Mr. Rufus H. Pope, Cookshire; Mr. M. C. Pearce, Stanstead, and of others.

Aberdeen-Angus—Shorthorn—West Highland cross-bred steer. His color was black, calved December 3rd, 1883. His weight in the middle of September, 1885, was 1,225 lbs., though not pushed at all. He was just a fine beast, plump and smooth. He was got by Northesk (Aberdeen-Angus) (1578), dam Bride, by Duke of Oxford 35th (Shorthorn) 2635c, g. d. Sheila (Shorthorn West Highland), by Sirius (Shorthorn) 18336, gr. g. d. Cruinach Og (West Highland cow) imported from Argyllshire, Scotland. A steer calf almost of like breeding was a compact animal, low, level, deep, and of fine quality, and took an easy first. While this latter cross may be of great service to ranchmen whose high latitudes require hardy beasts, the great lesson from all this to our countrymen is that for beef-production the Aberdeen-Angus upon our native cattle produces a most excellent cross.

J. J. Rodgers, Abingdon, Ill., has been one of the first to extensively use Angus bulls on to high bred Shorthorns. Estil & Elliot and Leonard Bros., two extensive Missouri firms, have also produced some excellent stock from native cows.

THE VERDICT.

So that we have not "far to go" for the verdict as to what is the best bull in the world to use for crossing or grading on beef cattle: "The nearer the animal is to the Aberdeen type," says Mr. Jas. Macdonald, "the better is he preferred by the butcher. If he is not pure on both sides, they try to get him sired by an Aberdeen bull."

CROSSES OF THE ABERDEENS ON THE DAIRY BREEDS.

After thus detailing the results of crossing the Angus on to cows of the beef breeds, we present,

THE BREED THAT BEATS THE RECORD. 59

lastly, a few proofs of their value for crossing on to cows of the milk breeds. One case has already been noted on the Ayrshire. Mr. Geo. Wilken, Waterside, Alford, Scotland, has conducted some experiments on the same breed.

In 1878 he bought three Ayrshire heifers, and served them with a polled bull. Next year he had three very pretty black polled heifer calves. He retained two of these, and had two calves every year from them—in all nine calves. All the produce have been black and polled, except one, which had a white spot on its side. The first cross retained the milking qualities, while they were quite superior beef animals. The second cross of the polled bull seemed to stamp his breed's characteristics so thoroughly that polled breeders, who were shown one of the produce among the herd, failed to point it out, though an expert cow dealer did. The writer was one of the number, and he can assert that there was no apparent mark to distinguish it from the others.*

In harvest time there was purchased, to give "hairst milk," at Bridgend, Alford, a Dutch-Holstein cow. She had been served by a polled bull, and shortly after arrival dropped a fine polled calf, with some white spots. Subsequent produce of this cow and calf were thoroughly black, and polled. The females proved excellent milkers, and of first-rate symmetry—a huge improvement on the original dam. There was a curious thing in connection with this cow. In the field, though horned, she was effectually kept out of the community of the other dairy cows; she dared not approach them. If she did, it was only to be chased away. It was a

---

*See also "Polled Cattle," by Macdonald & Sinclair, p. 389.

60   THE BREED THAT BEATS THE RECORD.

clear case, in this, of "the homyl beiring the wyte." The produce were greatly appreciated. The first calf out of the original cow was sold at the Bridgend displenish for a high figure. This cross cannot be too highly recommended.

Mr. Isaac B. Lutz, writing to the *Farmer's Review*, says the "Angus makes a good cross on the Jerseys, for farm purposes, as they will give from four to five gallons of milk per day. If properly fed for milk, the quality will nearly average with the average Jersey. The calves of the Angus and Jersey can be fattened at any age to sell, ready for butcher stuff." Such is the result of·a past ten years experience.

CHAPTER VII.

Early Maturity.

SIZE.

The opinions of the following authorities, acquainted with many different breeds, have been noted:

Jas. Bruce, of Ruthwell, Annan, Dumfrieshire, Scotland (live stock agent, frequently a judge, and author of "Scotch Live Stock"), says: "They grow to large size."

T. W. Harvey, Chicago, Secretary of the American Shorthorn Herd Brook Society, says: "They rank with the largest."

Hon. L. F. Allen, late editor of "American Shorthorn Herd-Book," author of various works, says: "They are equal to the largest."

Jas. Macdonald, editor of *The Live Stock Journal*, joint author of "History of Hereford Cattle," etc., says: "They are not inferior to the best Shorthorns or Herefords."

H. D. Adamson, winner of the Elkington Champion Plate, in 1879, at Birmingham, with a Shorthorn heifer, says: "Their color and compact contour deceives one very much—put the tape around them and they will astound you. Put them on the scales and you will be 200 pounds wrong on 1,000 pounds."

Robt. Bruce, Great Smeaton, Northallerton, England, well known on this side as one of the best judges in Britain, says: "It is their fine bone and no waste that at first sight might give an onlooker a wrong impression. I have never seen larger or better cattle in the show-ring than I have from this breed."

The following extract from *The North British Agriculturist*, says: " Bulls or oxen fattened for exhibition, scale as much as 2,700 lbs.; and we have seen females of the breed exceed 2,000 lbs.; a good average live weight for cows of the breed, as they go to the butcher, is from 1,200 to 1,400 lbs.

OFFALS.

In 1829, Mr. Hugh Watson's (Keillor), famous Smithfield medal heifer, had phenomenally fine bone—the bone of her foreleg was " no thicker than that of a deer." The tests prove that their offals are lighter than any other breed. That is best proved by the fact that they dress the highest *per cent.* of any. Black Prince, Geary's famous steer, I am told, by Mr. Geary himself, really dressed 71.50 per cent.; and the official record of J. J. Hill's Benholm, 1885, was 71.4 per cent.

Again, the bone is of the finest, and will compare favorably with all other breeds. I may give as an instance—a Newcastle butcher bought at the same time a Shorthorn and an Aberdeen heifer, and, for curiosity, had the hough bone of each neatly taken out and weighed. The carcass of the polled heifer weighed 784 lbs., that of the Shorthorn 448 lbs., yet the bone of the latter weighed 6¼ lbs., while that of the polled

THE BREED THAT BEATS THE RECORD. 63

heifer was only 5¾ lbs. Another instance—Mr. Bruce, the well known cattle salesman, of Newcastle, writes on the 9th of September, 1884: " Last year I sold a Sunderland butcher a black Polled-Aberdeen fat heifer, some two years seven months old, and the same day he bought a Shorthorn bullock, about the same age, weighing twenty-eight pounds more. The bone of the foreleg of the bullock weighed *exactly double* that of the heifer. The breed is a favorite equally with the consumer and butcher. With the former for its juicy flavor and quality of fat, and with the latter for the high percentage of dressed meat to live weight—67 to 70 per cent.—the flesh being so well intermixed with fat gives a mottled appearance like marble ; there is no waste from patchiness or blubber, the fat is well distributed, and the internal fat is of the finest quality."

The writer of a very important article in the *Scotsman*, in the fall of 1873, among other features, stated that "the offal of the Angus is remarkably light, lighter than any other Scotch breed, and eclipses the Shorthorn in evenness and plumpness of flesh. They do not consume so much food ; and are much hardier and finer in the bone than the Herefords."

DRESS OVER SEVENTY-TWO PER CENT. AT POISSY— 1857.

The following is quoted from *Le Fermier*, a French agricultural journal :

" At the Concours de Poissy, of 1857, much notice was taken of three Aberdeen oxen, aged from four to four-and-a-half years, having an average weight of 1,088 kilos. Bandement, who followed them to the butcher's, relates that they gave an average of 742 kilos. of meat, which would be 68 per cent., and 105 kilos. of tallow, which would be 9.6 per cent. But this average hides individual differences which it is interesting to recall. The oldest of the three gave a prodigious yield in net weight, surpassing 72 per cent. This was the highest of all ascertained yields at the Concours. This shows to what great weight the finest of this breed may attain, and what is their power of assimilation. I should add that at our last show of fat cattle at the Palais de l'Industrie, the highest yield of prime animals did not exceed 68 and 5-6th of beef."

EARLY MATURITY.

Henry Evershed, who is one of the best known writers in England, in an article on Sussex cattle, in the *Agricultural Gazette*, and who has published a work on Early Maturity, bears an independent testimony to the Aberdeens "as peculiarly noted for their early maturing properties." He was referring to the precocious maturity of the champion Paris group. It was not till 1881, however, that Shorthorn men, who had been in the habit of claiming this as their special excellence, laid down the gauntlet to all breeds. At the Smithfield Show, of 1881, rules were instituted that were meant, by the test of early maturity, to give a blow to what were dubbed "the slow maturing breeds." What was the result? That year the champion heifer and ox, and, consequently, championship of the entire show, were two Polled Aberdeens, *less* than two years and eight months old. These were exhibited by one man, and performed an unprecedented feat. and one that it will almost be impossible to rival again.

They were the youngest champions ever seen at Smithfield. That fact should never be forgotten.

At the Smithfleld Show, in 1879, the highest increase in weight per day, from birth, was shown by a two year and nine months old steer of the Polled Aberdeen or Angus breed. At Smithfield, in 1880, the average daily increase in weight of six Polled Aberdeen or Angus steers, under three years old, was 1.78 lbs., and that of the corresponding class of Shorthorn steers 1.79 lbs.

The following are the figures, from the *Mark Lane Express*, for Smithfield Show, 1880, in

CLASS 26.—Scotch Polled Steers, not exceeding three years old.

No.				
158.	Sir W. C. G. Cumming, Aberdeen,	1066	1892	1.77
160.	Sir W. C. G. Cumming, Aberdeen,	1023	1811	1.77
162.	R. C. Auld, Aberdeen.............	970	1876	1.93
163.	Robert Jardine, M.P., Galloway ..	954	1596	1.78
164.	Robert Jardine, M.P., Galloway ..	961	1534	1.60
165.	J. J. Colman, M.P., Aberdeen	1076	2072	1,93

On comparing these figures with the corresponding class of Shorthorns, we find only *one*, in that class of twelve entries, reaches 1.93.

The Agricultural Gazette, in commenting on the class alluding to the Galloways, said: "And two, *well, they are not fat at all.*" If it was not *beef* that gave this growth, it must have been the waste or bone or "timber"—as in the ox instanced above by Mr. Robt. Bruce, under the head of "offals."

In 1881—year of the Altyre victories—the animal making the highest average daily gain, of the cattle exhibited at the Islington Fat Stock Show, was the Scotch polled steer bred and exhibited by Mr. John Seaman Postle, of Smallburgh, Norwich. This animal

66 THE BREED THAT BEATS THE RECORD.

showed a record of 1,802 lbs. at 617 days, and an average daily gain of 2.91 lbs. He thus added his mite—not an unimportant one—to the heavy score made by the Scotch cattle at Islington.

The following table gives the weights of cattle of the breeds exhibited at the fat show in the Agricultural Hall, Islington, London, in 1881. It shows the number of animals in each class, their average age, their average gross weight, and their average daily gain in pounds:

STEERS NOT EXCEEDING TWO YEARS OLD.

Breeds.	No. of Animals in Class.	Average Age in Days.	Average Gross Weight in lbs.	Average Daily Gain in lbs.
Crosses....................	4	643	1458	2.26
Hereford............	12	581	1245	2.14
Shorthorn..................	6	639	1339	2.09
Sussex.....................	6	764	1394	2.06
Devon......................	12	662	1135	1.71

STEERS NOT EXCEEDING THREE YEARS OLD.

Shorthorn....	4	983	1975	2.01
Crosses.....................	12	907	1830	2.01
Scotch Polled	8	945	1874	1.99
Hereford....................	10	975	1734	1.79
Sussex......................	7	1022	1756	1.71
Norfolk and Suffolk Polled...	2	1058	1583	1.59
Devons.....................	10	1004	1465	1.45

STEERS NOT EXCEEDING FOUR YEARS OLD.

Crosses...	4	1358	2374	1.74
Scotch Polled................	2	1287	2106	1.73
Hereford......	9	1245	2002	1.60
Shorthorn...................	9	1413	2226	1.57
Sussex........	5	1310	2005	1.53
Norfolk and Suffolk Polled...	3	1267	1733	1.37
Devon.......................	7	1294	1602	1.24
Highland....................	4	1700	1853	1.00

HEIFERS NOT EXCEEDING FOUR YEARS OLD.

Crosses	8	1007	1629	1.61
Shorthorn	16	1196	1796	1.50
Scotch Polled	2	1197	1783	1.49
Sussex	5	1174	1721	1.46
Hereford	5	1115	1575	1.41
Devon	4	1181	1299	1.10
Highland	3	1371	1481	1.08

Referring to this memorable show, *The Field*, London, December, 1881, said: "It was clear that *early maturity and quality of meat weighed largely* with the judges in the awarding of the Champion Cup and other special prizes at London. Sir Wm. Gordon Cumming's champion heifer left nothing to be desired either in the quality of her meat, the fineness of her bone, or the symmetry of her form, and she also weighed well, showing a daily gain in weight, from her birth, of 1.85 lbs. * * * There was a good muster of Scotch breeds, more especially of that valuable and fast rising breed known as Polled Aberdeen-Angus. In average merit the last breed was excelled by none, and, as is well known to our readers, carried off the lion's share of the special honors."

This show proved the most successful for many years, having been visited by 124,683 persons. The remarkable success of the Aberdeen or Angus Polled cattle was the subject of comment by the press of the three kingdoms. Scotch cattle were never so successful before at Smithfield, the remarkable feature being the early maturity of the two animals belonging to Sir Wm. Gordon Cumming. The champion heifer, for her age of 2 years 8 months, showed a weight of 15 cwts. 3 qrs. 24 lbs. The best heifer of the Hereford breed was four months older, but 1 cwt. 2 qrs. 8 lbs. lighter;

and the best Devon heifer, also four months older, but 3 cwts. less weight. The best Shorthorn heifer was 4 years 10 months 7 days old, and weighed 19 cwts. 2 qrs. 7 lbs., showing scarcely 4 cwts. for two years and two months extra growth. The polled ox of Sir Wm. Gordon Cumming, which took the prize as the best ox of any breed, had also greatly the advantage in weight for his age. He was 2 years 8 months and 6 days old, and turned on the scales 17 cwts. 1 qr. 21 lbs. The Hereford ox of Mr. Lloyd stood next in favor with the judges, but he was three months older and 1 cwt. 1 qr. lighter. The best Devon ox was four months older than the polled and $2\frac{1}{4}$ cwts. lighter. The best Shorthorn ox, Mr. R. Wortley's (bred at Craigwillie), was two months older than the Polled, but one half hundredweight lighter. (The English "cwt." is 112 lbs.)

The demand for young beef is only of late origin. Mr. H. D. Adamson put the matter correctly when he said, "formerly the Scotch grazier thought it necessary to send nothing to the London markets under four years of age, and the national fat stock shows gave no encouragement to early maturity." It was, indeed, rather a hobby with breeders of the older school to keep their cattle till they were aged; and, further, it was thought and agreed, and maintained and held, that beef could not be healthy or wholesome till the ox was matured and mellowed by age. That was the secret of the matter. Young beef it was thought could not be wholesome, could not contain in its tissue the elements that made it food for the strong. National markets and shows, indeed, encouraged and supported the idea. But a new order of things—the necessity for a quicker return of capital—led breeders and feeders to discover that early maturity did not mean unwholesome or unmellowed beef. The early maturity step was, therefore, taken, and it has come to be the chief test of breeds, and the chief point to be encouraged. The transition from the four-year-old champion polls of 1857, 1862, and the three-year-old champions of 1872 to the champions of 1881, and the present day, is complete. Under the new test, the Aberdeen-Angus at once proved their powers. In 1883 the ten young Aberdeen steers averaged 1,804 lbs., the Herefords, 1,742 lbs. Mr. Stevenson's Aberdeen polled heifer, which was champion at Birmingham, in 1883, at the age of two years and eight months, scaled 1,867 lbs., whilst the Hereford heifer, which was awarded the cup for the best of its breed at the age of *three years six months*, scaled only 1,615 lbs.

The following figures relate to the (1883) Smithfield winners of the " Breed Cups " :

	Age in days.	Daily gain in lbs
The Queen's Shorthorn heifer........	987	2.08
Mr. C. Stephenson's Polled heifer....	984	2.01
Mr. J. S. Hodgson's Sussex steer.....	1083	1.97
Mr. I. Myddleton's Hereford steer....	1272	1.67
Mr. J. Baker's cross-bred ox.........	1359	1.63
Major Platt's Welsh ox..............	1416	1.63
Prince of Wales' Devon steer........	1397	1.34

(Norfolk Polls and Welsh compete together.)

The first on the list was the champion of the show. A report said: " Mr. Stephenson had another heifer, which was champion at Birmingham, which could not compete at Smithfield. We should have liked to have seen her try conclusions here on this occasion." (See p. 52.)

The 1885 Smithfield Show had, for the first time, a class of young Polled Aberdeen steers, not exceeding two years old. The following are the figures, and comments from *The Field* which are valuable for comparison and reference:

"It is now demonstrated that the yonng classes are those in which meat is put on the most cheaply. The vigorous digestive powers of the growing animal extract more of value from the food given to it than any older beast can be hoped to return, Now, besides the question of the age at which meat is most profitably made, comes up a hardly less important question—'By which breed is the early profitable beef most freely laid on?' Here are a few statistics culled from the catalogue of the Smithfield exhibition of 1885, which is a kind of model in its way. The youngest classes only are taken, and those breeds only are dealt with which lend themselves the more readily to early maturity:

	Days.				Lbs.
7 Devons at an average of	627—	Gave average weight of			1158
5 Herefords "	597	"	"	"	1368
8 Shorthorns "	668	"	"	"	1553
8 Sussex "	688	"	"	"	1468
7 Polled Angus "	604	"	"	"	1408
11 Cross-breds "	605	"	"	"	1800

"If these ages and weights be reduced so as to show what each animal of each variety had, on an average, to show for each day of its life, it will appear that

		Lbs.
Each Devon for each day had	....	1.85
Each Hereford "	....	2.28
Each Shorthorn "	....	2.48
Each Sussex "	....	2.14
Each Polled Angus "	. ...	2.49
Each cross-bred "	....	2.28

"These variations are sufficiently surprising, and point to truths not always recognized.

"The largest rate per day is made by the black polled

breed, which actually makes a larger growth per diem than does the same poll when a cross with the quickest ripening blood [?] is admittedly present."

At the Kansas City Fat Stock Show of 1885, there were exhibited of Aberdeen-Angus—(1), in the two-year-old class, "Blaine" and "Logan;" (2), in the yearling class, "Sandy;" (3), in the calf class, "Alex."

In the early maturity class these came in body bulk ahead of all others:

Two-year-old class...Blaine..weighing 1615 lbs., age 761 days.
Two-year-old " .. Logan.. "   1520 "   " 651 "
One-year-old " ...Sandy.. "   1455 "   " 581 "
Calf   " ...Alex.... "   905 "   " 292 "

These tests place the Aberdeens and Shorthorns, and their combinations, before all other breeds. This is the remarkable result of the world's test. With it we are satisfied.

CHAPTER VIII.

Prime Scots.

AS BEEF PRODUCERS.

From the earliest time the Aberdeen-Angus cattle have been famed for the quality of their beef. In the 4th century the Buchan cattle supplied beef to the Roman legionaries. It was for their qualities in this respect that they were for centuries sought out. Dr. Skene Keith says, "that no place in the kingdom could boast of better beef than Aberdeen." It is described as finely grained and marbled—the fat and lean equally distributed, juicy, tender, and as well flavored as the Kyloe. These characters have been retained, and Aberdeen, to-day, can still boast the finest meat market in Britain. Their fame early reached London. Rev. A. Forsyth, LL.D., of Belhelvie, wrote in 1837: "A great many cattle are bred and fed in this parish for the London market. They are principally of the improved Aberdeenshire breed. Their bones are small, they carry a great deal of flesh, are easily fed, and are soon fit for the market." Rev. Chas. Gibbon, of Lonmay, writing in 1830, says: "It is well known that Buchan has long been celebrated for its cattle. For the last twenty years, preceding 1830, the polled or doddied were in great demand, and, indeed, still bring high prices in the southern markets and *the top price in London.*" In the Chapel of Garioch district, about the same time, "during the winter, a great many are fed on *turnips and straw*, and are either sold to the butcher, or sent by sea to London." The oil-cake cross had not yet been introduced. This beef character has been since annually proved to demonstration, at the London market—the best market ,in the world for prime butcher's beasts.* And here it is necessary to say that Polled Scots, or Prime Scots, mean the Polled Aberdeen-Angus. This should be distinctly understood, once for all. Speaking of the breed, the *Farmer's Review*, which has the best live stock department of any agricultural paper in America, says of the breed: " Their beef is noted for its excellence, and is called 'Prime Scots' in the Smithfield, London, market, where it brings two cents more than any other beef." This, in fact, is written by the editor, a gentleman hailing from the southwest of Scotland.

Again, the *Mark Lane Express*, in 1879, says: " The great value of the Polled Scotch breed of cattle, as beef producers, is well known all over England, and especially in the London markets. At the Paris exhibition, last year, the gold medal offered for the best beef producing breed was awarded to the Polled Scots. [This sentence is given to show the identity of the Polled Scots with the Aberdeen Polled—which can be so much fur-

---

*It is of great interest to note, in connection with the beef trade, that these two great industries, that have reached such dimensions in America, viz., the canning and dead meat trade, originated first in the city of Aberdeen, which exclusively conducted both for many years.

ther illustrated]. Their beef is of the very best and they are exceedingly fine in the bone, and light of offal; standing, in this respect, far above the Shorthorns."

It is the same journal, the *Mark Lane Express*, in reviewing a work on "Meat Production," by John Ewart, in which the Aberdeens are placed second, and the Shorthorns and Hereford first, says: "Here again the author is in error as to facts; *for an Aberdeen bullock is thicker than either the Hereford or the Shorthorn.*"

The Country Gentleman says: "'Prime Aberdeen' is a phrase which is in common use in the London beef market. Perhaps if you were to ask every butcher in London, you would find he never sold anything else."

Mr. Hine has visited the London market and finds the Aberdeens are the Prime Scots: "The flesh is finely marbled, and the best portions of the carcass are well developed. The amount of offal is about ten per cent. less than in most other beef breeds. The London, England, market recognizes the superiority of the Aberdeen-Angus beef in that the 'Scots' are held for the holiday sales, and command two cents per pound more than other beef. The patrons of other breeds are not unwilling to accord to the Aberdeen-Angus the first place, from a beef-producing standpoint."

Col. G. W. Henry, of Angus Park, Kansas City, in correcting the eggregious but perennial error of asserting against the absolute knowledge to the contrary, that the Galloways contribute to the "Prime" Scots, says; "I would like to know a little more about the London market he (Mr. McCrae) speaks of. I spent a month in the London markets just four years ago, the state of facts he speaks of did not then exist, and possibly he is speaking of more recent dates."

G. L. Stichter, Kenesaw, Nebraska, noticing the same allegation, writes also to the Kansas City *Live Stock Indicator*, March 11, 1886: "His (Mr. McCrae's) assertion that 'in the London market Galloway beef sells one penny per pound higher than other cattle' *must certainly have originated in a dream.*"

The special reports in the London *Live Stock Journal; North British Agriculturist*, Edinburgh; *Breeder's Gazette ; National Live Stock Journal;* Kansas City *Live Stock Record;* of the London market, for 1885, *all* say that "no Galloways were there "—that it was the Polled Aberdeen-Angus that were the " Prime Scots," as they always have been.

In *The Live Stock of the Farm*, edited by that veteran, John Chalmers Morton (Bradbury, Agnew & Co., 9 Bouverie Street, London), 1882, says of the Aberdeen-Angus, p. 24: " Pure or crossed with the Shorthorn, sent to London, either as live animals or dead meat, command the top price of the market."

The following is from the London *Daily Telegraph*'s report of the Christmas market for 1876: " The principal feature of to-day's show was the receipt of Scotch cattle. About 2,000 head came to hand. We believe we are correct in stating that during the past week fully £40,000 worth of stock left Aberdeen, alone, for the London market. This breed, undoubtedly, formed the *piece de resistance.* Take away the Scots and the principal charm disappears—in fact, they may be considered to have chiefly redeemed the market. There was also, of course, a good show of cross-bred, with a fair sprinkling of Devons and some prime Herefords."

Mr. H. D. Adamson says: " Take up an English agricultural paper, or even the *Times*, refer to the London fat cattle markets, and you will at all times find the Scots, *that is the Aberdeen and Angus*, or their crosses, commanding the highest figures, on an average of two cents per pound more than the Shorthorn, for the carcass dressed. Why is this? Because not only is the meat better, but the butcher knows he has much less waste,"

Mr. J. L. Thompson, manager at Beefacres, South Australia, in a paper read before the Royal Agricultural Society of that colony, alluded to the pre-eminence of the Aberdeens: " Where it is intended to ship or 'train' (car) the cattle to a distant market, alive, I would recommend the Polled Aberdeen. There is no denying the fact that, for beef, the Polled Aberdeen commands the highest price in the London market. Take up any English paper, and cast your eyes to the meat reports, you will always find 'Prime Scotch' quoted, say, at 5s. 10d. to 6s. per stone of 8 lbs.; 'Hereford,' 5s. 8d. to 5s. 10d., etc."

The following is the usual way the market is reported:

"Prime Scots...............	6s.	to 6s. 2d. per stone.
"Shorthorns, Herefords.....	5s. 8d.	to 5s. 10d. " "
"Inferior Beasts............	5s. 6d.	" "
"American, etc............ .	5s. 4d.	to 5s. 6d. " "

" The top quotations mean Scots cattle, and Scots only." And now it is a common thing to see in the reports of the market special reference to the Polled Aberdeens—the Prime Scots: "Prime Scots made 6s. 2d., and, in some cases, best Polled Aberdeens, 6s. 4d." (*Live Stock Journal*, December, 1880; see also 1884). So that each Christmas market is a recurrence of the old demonstration. "The Christmas market, like every other one that has gone before, simply proves that it is impossible to obtain a better feeding, heavier weighing, or a more profitable bullock than the Polled Aberdeen." (Special report *Daily Free Press*, Aberdeen, December 13, 1881.)

"The typical Aberdeen is an animal with a fourth less bone and offal than an average Shorthorn, carrying lean of a far finer quality. It is such close making up that makes it weigh so much in proportion to its size, and in comparison with other breeds." (Same report.)

Mr. Wm. Anderson, Wellhouse, Alford, an unimpeachable witness, says: "In my experience the Polled Scot* is the best selling animal in good times, and the best selling animal in bad times. As a rule, I get £2 ($10) a head, or even more, for polled animals than for crosses of the same weight * * * the butcher can well enough afford that extra sum. I lately heard the statement of a leading Aberdeen butcher, that he could give 5s. (about $1.25) more per hundredweight (112 lbs.) for a fat polled animal than for a cross."

One would not expect the *Quarterly Review* to contain much allusion to beef, bucolics, or butchers. Nor does it; but when it did it would be all the more noteworthy. The *Quarterly Review*, about 1850, published an article by the late Mr. Thomas Gisborne, M.P., of Yoxall, Staffordshire, a man of the time and known to fame, who contributed many articles on agriculture, collected and published about 1857. In that on "Cattle and Sheep," he thus places the breeds for beef—
"Scots generally, Devons, Herefords, indiscriminate crosses and mongrels, down to the improved Shorthorns. In each case the butchers' shops will confirm our lists. There the animal that stands at the top will

---

*Again this term is applied solely to Aberdeen-Angus, by a breeder.

sell for at least 1d. per pound more than that which stands at the bottom." Than that nothing could be more decisive; and it is conclusive.

The London *Standard*, on the Christmas show of 1883, said: "We are not afraid of American competition driving the English farmer out of his native land, *so long as he* can produce *Aberdeen beef*, such as no pasture except ours grows, and Welsh mutton, which most prefer to venison, he need have little dread for the pasture." "The Roast Beef of Old England!"—that has long given place to and been " Prime Aberdeen."

A clipping from the London *Times*, of Tuesday, December 31, 1878, says— and the reader will now be able to appreciate its significance: " Last week's value has been maintained, but it has been with difficulty; six shillings per stone means for Scotch cattle, and Scotch only. Quotations of the day—Prime Scotch, 6s. per stone; Herefords, 5s. 10d.; Norfolks, 5s. 8d. to 5s. 10d.; Lincolns (best) 5s. 8d. to 5s. 10d.; Americans, 5s. 4d. to, in some instances, 5s. 6d. per stone."

Mark Lane Express, October 21, 1878: " The top price for best Scots was 5s. 10d. per 8 lbs.; many good Shorthorns did not make more than 5s. to 5s. 4d. per 8 lbs." The relative value of Shorthorns to other breeds is expressed in the same ratio in every other market returns of which I have any knowledge.

Mr. G. T. Turner, in the *National Live Stock Journal*, 1878, says: "The Polled Scots are all the year round beasts; are well fed and full of prime flesh in prime parts. They usually top the market, and for thick cutting and value for money, at all times, they certainly bear the palm."

The *Live Stock Journal*, London, 1880, report, says: " As to quality, the Scotch breeds occupied pre-eminently the first place this year at all our fat stock shows. Scotch cattle are put above all other breeds by the London, butcher. Of the Polled Aberdeen he has the highest opinion. The late Mr. McCombie, of Tillyfour, who made and maintained the reputation of the celebrated breed, was excellently represented by a large consignment bearing his well known (rump) mark, M X C. The general quotation may be put at 6s. per stone of 8 lbs.; Prime Scots made 6s. 2d., and, in some cases of the best Polled Aberdeens, 6s. 4d."

Mr. McCombie, at a Rothiemay sale, made the following memorable remarks: " We are deeply indebted to our friend and to those noblemen and gentlemen who have espoused the welfare of the country by their determination to keep our native breed of polled cattle pure, without the mixture of Shorthorns. I admit that Shorthorns are useful for crossing; I admit that a cross between a Shorthorn bull and a polled cow is a good animal; but I deny that their qualifications go farther. But if the pure breed is lost, what will your crosses become? Look at a neighboring district where they had, since I recollect, the finest breed of polled cattle. They introduced Shorthorns; they have crossed and recrossed. Their cattle are white instead of black; they are only the shadows of their former greatness—(laughter). The pole-axe is the end of all breeds. Our polled breed stands at the top of beef producers, the Shorthorns at the bottom. At the great International Show at Poissy, all the different breeds were measured and weighed. The polled were found the heaviest of all breeds by their measurement, the Shorthorns the lightest of all breeds—(laughter). London is the greatest fat market in Britain, and, as a class, the westend butchers are the greatest and the wealthiest in the world. What is the great sensation on the great market day? Is it the long dark lines of our polled cattle, or is it the long lines of our white and red Shorthorns? On the morning of the great day the onlooker will find the west-end butchers, with the best of the country butchers, congregated behind and before the black lines, and the lines of Shorthorns almost deserted, except by a few carcass butchers. Ask the west-end butcher if the Shorthorn suits his trade? Ask him how much more he will give for a Polled Scot than for a Shorthorn per stone? Ask him what value he puts upon the descendant of a £4,000 Duchess?"

AUTOBIOGRAPHY OF A "PRIME SCOT."

From an article, written by the "Druid," in *All the Year Round*, December 21, 1872, entitled "A Prime Scot"—an autobiography, from which we are fain to make two extracts, the first and last paragraphs :

Moriturus vos saluto ! To-morrow morning the pole-axe will sink into my forehead ; the day after, my prime joints will be exhibited in all the toothsome streaky splendor of fat and lean in the shop of that fortunate west-end butcher who "clapped hands" for me at the Great Christmas Cattle Market. It is my fate to be eaten, but no vulgar tooth shall masticate my firm yet succulent tissues ; indeed, I have reason to believe that I am all bespoken already. I die happy, my mission and my ambition have been fulfilled. My worst enemy, if I have one, cannot but say, in the language of Peter Allardyce, that "I fill a string well." With three comrades, 1 had the proud position of topping the Christmas Cattle Market ; in the racy language of the lamented John Benzies, I am "beef to the root of the lug," and I have a well-grounded conviction that I shall "die well" in the sense in which butchers use the term. True, the immortality of being exhibited at the Great Smithfield Club Show has been denied me owing to an unfortunate congenital lack of perfection in the region of the hook bones and a trifling defect of symmetry behind the shoulder; but as a rational ox I cannot grumble at the decree of fate, and it is something, surely, to have topped the Great Christmas Cattle Market. It has not been without willing exertions on my part, and incessant attention on the part of those who have had the charge of me from calfhood till the day I left my home "prime fat," that this distinction has been achieved. I have my reward in the proud consciousness of that distinction, while my breeder and feeder has his in the long price that was given for me without a murmur.

I am a three-year-old. I was born at Tillyfour, the abiding place of the "powerful, pushing, and prosperous race" of McCombie. In my veins runs the best blood of the breed to which I belong, the Black Polled Aberdeenshire. My genealogy goes back to the famous old Queen Mother, the corner-stone of the Tillyfour fortunes in the polls. My destiny to die fat was fixed from the very hour of my birth, and that I should fulfill my destiny a settled scheme—the result of long experience and intelligent observation—was sedulously pursued.

After fully describing the incidents of the fatting process till he was ready to ship to London, he describes his arrival there and the closing hours of his life:

From the Maiden-lane station we were directly driven off to the lairs that had been secured for us in the immediate neighborhood of the cattle market, and spent a pleasant and grateful night in the midst of clean straw and plenty of food. Sunday was a day of profound and welcome rest, the only incident of which was the clipping of the initial "McC." in the long hair on the flank of each of us. Long before dawn on the Monday morning, the morning of the great Christmas market, we were on our way from the lairs to Mr. Giblet's stances hard by the western side of the great tower in Copenhagen Fields. It was not long after daybreak when the swell butchers came thronging around us, pinching, poking, and praising. Mr. Giblet, who acts as salesman for Tillyfour, said very little, but stood by with an assumption of unconcern—"We were beasts," I had heard him say, "that would sell themselves." I was sold, with three others, before nine o'clock, and, as I have said, we topped the market. Driven back to the lair, I now serenely abide there the blow of the pole-axe, conscious that I have deserved well of my country, that there is not a coarse bit of meat about me, that I shall be found to have surprisingly little offal, and that after I am gone it will lie in the mouth of no man to give me a bad name.

"PUNCH" ON POLLED SCOTS.

"The only illustrated history of our own times"—London *Punch*—has on more than one occasion heard of the fame of the Aberdeen-Angus cattle, and we quote one of the "jokes of the day" from that journal: "Different Droves.—At recent live meat shows much attention has been attracted by some particularly fine specimens of Polled Scotch cattle. Polled howsoever these cattle may be, they are a breed incapable of bribery."

CHAPTER IX.

The Epicure's Choice.

THE BEEF.

The *National Live Stock Journal*, June, 1881—
"Angus Cattle to the Fore"—says: "The beef of this choice breed brings one to two cents more per pound, in the London market, than the best of any English breed; and the bullocks can be reared at least 10 per cent. cheaper than horned cattle. Indeed, some who have kept polled cattle alongside of horned, both in Great Britain and America, say the cost of rearing them for a beef market is 20 to even 25 per cent. in their favor."

The French journal, *Le Fermier*, speaking of the test of the beef by the order of the Emperor, at Paris, in 1862, says: "All the characteristics which denote a great aptitude for fattening are combined with those which indicate a considerable living weight and an eminent yield in solid beef. The muscles are everywhere equally developed, compact, and firm, well marbled with fat where fat is suitable. Upon all the dorsal regions, especially, the Aberdeen take on a thickness which gives a great value to animals in a country where roast beef ("ros bif") and rump steak are *recherches*. The flesh of the Aberdeen is of an exquisite flavor, and in the London markets a considerable higher price than for that of the English or other breeds is obtained. The fat, which is stretched in a thick equal covering under the skin, or deposited between the muscular masses, is of itself a close, fine tissue full of taste and flavor."

TESTED IN LONDON—IT IS "THE BEAU IDEAL OF COOKS."

The Agricultural Gazette, February 24, 1874, at that time the leading Shorthorn organ, had some articles on Beef. The writer, in concluding, said: "As a further instance of the way this question has been tested by the Smithfield Club, I may add that I have just been informed that Mr. Giblet had a joint on a side table in the council room, at the time of the last show, which he brought there as a specimen of the meat which a prize animal ought to supply." This roast was from a two-year steer, consigned the week previously to the London market, by Mr. McCombie, of Tillyfour, who received from Mr. Robt. Leeds, chairman of the Agricultural Hall Company, a letter of thanks, in which he said: "I wish particularly to thank you for the very best piece of beef we ever had on our table during the cattle show." Who could be better judges? Last year, at Chicago, were exhibited the polled ox, Black Prince, and the famous Shorthorn, Clarence Kirklivington. Both of these came to the block, and their meat was tested afterwards by the best judges in Chicago. We have not sought the answer "in the stars" as to which was the "best," that is easily imagined.

THE BREED THAT BEATS THE RECORD. 85

We are able to give to the eye a representation of what Polled Aberdeen—Prime Aberdeen or Prime Scots—beef is like. The following remarks are by that excellent authority Mr. George Hendry, the agricultural editor of the *Free Press*, from that paper, commenting on the original photograph:

It is not as dairy cattle but as beef producers that the Polled Aberdeen or Angus breed of cattle have come so prominently to the front. It is not on account of their milking properties, though these we don't disparage, that American purchasers come all the way to this country for our blackskins. They come here because they have had the shrewdness to perceive that of all the breeds of cattle the polled is the nearest approach to a *beau ideal* beef producer. Carrying an unusually large proportion of beef to its bone, with little offal and skink bone, the polled ox or heifer commands a higher price per pound, say from 2s. to 2s 6d. per cwt., in the home market than any other animal of a different breed. In addition to this, it carries its "meat" on the most valuable parts, and above all, the quality of the flesh of a pure-bred poll is, by connoisseurs, considered to be unequalled by that of any other breed. In order to show his countrymen how the Polled Aberdeens "cut up," Mr. A. B. Matthews, Kansas, has obtained a photograph of the "meat" of a two-year-old pure-bred polled ox, fed in the Vale of Alford, and which was exposed in the shop of Mr. John Williamson, jr., New Market, Aberdeen. The photograph, which was executed by Mr. G. W. Wilson, shows cuts from different parts of the animal—the shoulder, the rump, the steak piece, and the "Pope's Eye," and the beautifully mixed flesh is well brought out. We have never seen beef in which the fat and flesh were so well mixed, and with the grain so perfect, and it speaks volumes for the breed that this could have been produced in an animal only two years old. Mr. James Martin, of Messrs. J. & W. Martin, and Mr. James Williamson, whose long experience in the cattle trade enables them to speak with authority on this subject, say they never saw such a grandly-fleshed animal in all their lives. It may be mentioned that he was picked up in one of the Alford cattle markets for Mr. Williamson, jr. Could the photograph of which we speak be reproduced in some of the many excellently illustrated agricultural newspapers of America, it would, we doubt not, tend still further to increase the prestige of this valuable breed of cattle on the other side of the Atlantic. So long as breeders of polled cattle can produce stock of this description, the proud position which the polled breed occupies will be assured.

The *Breeder's Gazette*, in 1882 (July 27), also gave an engraving of this, and remarked: "It has seldom been our fortune to see a more perfect admixture of fat and lean in the flesh of any animal." It is this intimate admixture of fat between the fibres of the meat that constitutes the highest *acme* of beef perfection.

IN CHICAGO.

Mr. T. W. Harvey thus descants on Aberdeen beef: "In London the Aberdeen-Angus is quoted at the top of the market, and at Christmas time it brings two cents above the market. One of the great annual events in Aberdeenshire is the shipment of car loads of these cattle by special trains to the cities, to meet the demand for choice beef for the holiday festivities, and we understand that it is an Aberdeen-Angus roast that the Queen partakes of on Christmas day. In Chicago, at the last Fat Stock Show, two Aberdeen-Angus cattle were killed, a two-year-old steer and a cow. Both were exhibited by the Hon. H. M. Cochrane, of Canada. The fat, instead of being in great rolls or layers on the outer part of the back and down the sides, causing waste and excess to both butcher and consumer, was very delicately marbled through the lean, producing the most delicious meat, and was pronounced by some of our city epicures as the finest beef they had ever eaten. The large three-year-old Aberdeen-Angus steer, that was awarded sweep stakes by the butchers, in three-year-old class, at the same show, was not killed, but will be kept for next year. A well known business man of Chicago, who thoroughly appreciates what is excellent, was so enthusiastic after tasting of the meat of Mr. Cochrane's steer, that he declared himself willing to pay forty cents a pound the year round for such beef. There certainly is a delicacy and lusciousness about the Aberdeen-Angus beef that is remarkable; it is beautifully laid in, and very tender.

Geary Bros. wrote to *The Gazette*, March, 1885, of their steer " Black Prince ":

Thh price per pound which we received for Black Prince was twelve cents. We sold him to N. T. Croxon, of the Union Stock Yards Restaurant, and in a letter received from Croxon, under date of December 25, he says : "Black Prince, as you suppose, is all eaten, with the exception of some very fine corned beef, in barrel, which is disappearing very fast, and is delicious. The first of him I served up cold to the delegates of the St. Louis Stockmen's Convention, commonly called the 'cow boys,' in which were represented the heaviest cattle interests of the country, and they pronounced the beef to be very fine, as indeed it was, and I can answer for it myself." Black Prince was sold on his merits and without any pressure being brought to bear, as was the case with his more successful rival, Clarence, and no part of him was ever returned as being unfit for use, as in the case of Clarence, according to report. If it comes to a point as between the value of the two carcasses from a butcher's or consumer's point of view, we can't understand upon what the committee based their decision.

TESTED AT KANSAS CITY.

" They do things thoroughly in America. The result of the competition for carcasses of fat cattle at Kansas Fat Stock Show, last year, resulted in a victory over all breeds for the Polled Aberdeen-Angus cow Pride 3rd of Blairshinnoch 4658. A portion of the sirloin was sent along with a corresponding portion from the sirloin of a prize Galloway grade steer, as a gift to the editor and business manager of the *Kansas City Live Stock Journal.* These gentlemen had the samples of meat both cooked alike. A party of six gentlemen were invited to the table at which the meat was served, and on tasting it, five, including the host, pronounced in favor of the Aberdeen-Angus cow against one for the Galloway grade. Rump steaks from the carcasses were served at a restaurant, and five persons served with it, of whom four pronounced for the Aberdeen-Angus beef. 'To sum up the result—Both samples of beef were exceptionally choice and sweet, and the points of difference may be briefly stated: Aberdeen four-year-old cow—Flesh well covered with fat; lean meat, dark colored, and both fat and lean juicy, mellow and tender; somewhat coarse grained; flavor excellent. Grade Galloway two-year-old steer—Flesh moderately well covered with fat; lean meat light colored; fat more gristly, and lean meat not so tender on outside; close grained; flavor gamey.'"

At the late Chicago Fat Stock Show, Mr. P. D. Armour, of Chicago, purchased the carcasses of Hutcheon, Kinloss, and Turiff, for distributing the beef among his friends—a no mean compliment to the Angus.

At the Chicago Fat Stock Show much interest attaches to the result of the competition for the best dressed carcasses. The animals entered are followed "to the block" and a tabulated statement made of the yield of meat and offal. This year forty animals of various breeds and ages were slaughtered and dressed. In the competition the Polled Aberdeen-Angus two-year-old ox Benholm, belonging to Mr. J. J. Hill, gives

ABERDEEN BEEF. (See p. 85.)

the best result, as will be seen in next chapter. The ox was bred by Mr. Smith, of Benholm Castle, Scotland. The *Breeder's Gazette* said: "Mr. J. J. Hill's pure bred Aberdeen-Angus bullock Benholm's carcass was served by Mr. F. T. Croxson, at the Exchange restaurant, Union Stock Yards. A *Gazette* reporter dropped into the Exchange Buildings a few days since and caught Superintendent G. T. Williams just as he was picking his teeth after dining off a roast from the doddie. 'No man alive ever ate a better piece of beef,' was his reply to a query as to how the animal 'panned out' on the table. This declaration pleased Mr. Croxson very much, as he had selected Benholm as his choice after an extended examination of the entire lot, and now that he had tested him from the consumer's stand-point, was not at all disappointed. He pronounced him as fine as could be grown, excelling even Black Prince, which he had purchased last year, in that he was not over-ripe, but was just 'done to perfection.'"

Mr. John B. Sherman, President of the Stock Yards, remarked, with a smack of the lips, that he would "have to dine off Benholm for a week before he would pass his opinion on him," which may be construed to mean that the flavor of the beef tickled his palate not a little. Again it was "40c. a pound all the year round!"

CHAPTER X.

" Tuning Up."

THE POLLS AT AMERICAN FAT STOCK SHOWS.

Aberdeen-Angus cattle have appeared but very sparingly as yet at the American fat stock shows. No one could expect it to have been otherwise. But the exhibits have been of a kind to " naturally furnish the breeders with plenty of thunder "—as the *Farmer's Review* declared in 1884. We certainly expect lots of *music*. Just now we are only *tuning up*.

As the system in America is, these fat stock shows make each animal exhibited an object lesson of the quality of the breed from which it is drawn; so we have concluded to give the individual record of these Polls in this chapter. We think the best method is to treat of each year by itself.

1883.

KANSAS CITY.

At this show the following Aberdeen-Angus animals appeared: Geary's "Black Prince;" Col. G. W. Henry's "Bride;" Gudgell & Simpson's "Bruce's Queen;" Mathews' "Paris Heifer." The two former were shown in one class, the two latter in another, and they stood in the order given. The class was open to Galloways. "Black Prince" took the breed championship, also second against all in three-year-old early maturity class. "Black Prince" and "Bride" will come under notice again. "Bruce's Queen" and "Paris Heifer" are still alive and "doing well."

CHICAGO.

At this show the Aberdeen-Angus present were: Geary's three-year-old "Black Prince;" M. H. Cochrane's two-year-old "Waterside Jock," and cow, over three-years-old, "Duchess 2nd." There was thus only one in each of these classes, and they "got the premiums first and were admired afterwards." "Black Prince" was breed champion. "Waterside Jock" took the $50 premium as best two-year old dressed carcass. This animal was imported, for the purpose of exhibition at this show, from Scotland, where he had obtained some minor breed prizes.

Said the *Breeder's Gazette*, October 30, 1884, in an article "Dressed Carcasses at the Fat Stock Show:" "This animal attracted much attention, and was a good advertisement for the breed both on foot and on the hooks. He dressed the largest per cent. of net carcass to gross weight of any two-year-old steer awarded a first prize in this ring."

Year.	Breed.	Live Weight.	Dressed Weight.	Per cent. of Dressed to Live Weight.	Per cent. of profitable to Live Weight.
1883	Aberdeen-Angus	1,785	1,203	.67	.81
1882	Grade Hereford	1,735	1,115	.64	.76
1881	Grade Shorthorn	1,460	977	.66	.78

The following is the report of the judging committee on "Waterside Jock:" "The steer was very ripe, considering the age. The distribution of meat in the best parts of the carcass leaves no room to doubt that he would cut an unusually large proportion of net to gross. This steer was near perfection in all that goes to make up a profitable butcher's beast, thickly covered with the best quality of firm, mellow, and well-marbled flesh."

The cow class was open to all breeds. There were three entries. The prize went to "Duchess 2nd," the committee reporting as follows: "The first premium was awarded a cow that nearly approached the highest standard of perfection for a butcher's beast. Compact and blocky; fine in bone; small head and short, thin neck; thickly covered with firm, mellow flesh; quarters nicely proportioned and well meated down to hock and gambrel joint."

"Black Prince" and "Jock" occupied the class "other pure breeds not named"—there being yet no regular class for Aberdeen-Angus. The class report on "Black Prince" was as follows: "This grand specimen was a compact, blocky steer, with straight top, bottom, and side lines, and, considering the weight, was fine in bone, and there was but slight room for improvement. The steer was thickly covered with firm, mellow flesh, of that extra quality to be found in prime steers ripe for the block."

This animal, as noted, took the breed sweepstakes. He also took the $50 sweepstakes for best three-year-old steer or spayed heifer, "to be judged by butchers." The committee reported on this competition as follows: "The above ring, at the present show, was composed of twenty animals of superior merit, representing all the leading beef breeds and their grades. Considering the difference in breeds and the methods of feeding and handling, the cattle were quite uniform in form and quality. The cattle were fatted and in prime condition for slaughter. The work of deciding as to the respective merits of such a superior lot of cattle was a laborious undertaking, where, with scarcely an exception, all the animals nearly approached the standard of a perfect butcher's bullock. The sweepstakes premium was awarded to the Polled Angus steer, " Black Prince," exhibited by Geary Brothers, of London, Ont. This steer is a very superior specimen of the Breed, and could be but little improved in all that goes to make up a profitable beast for the butcher, and a desirable carcass for the consumer. The superior handling qualities of this steer gave every consumer assurance of a very superior quality of meat. The steer, considering weight, was fine in bone, and would cut a large proportion of net to gross. The top, bottom, and side lines could be but little improved. This steer was heavily and evenly quartered, and thickly and evenly covered in the most valuable portions of the carcass with firm, mellow meat."

1884.

KANSAS CITY.

The following Aberdeen-Angus appeared: From Indiana Blooded-Stock Co., " Blaine " and " Logan " (one-year-old); Col. Henry's " Bride." There was one Aberdeen-Angus grade— " Abernethy," from the American Aberdeen-Angus Breeders Association, Independence, Mo. And there was one cross-bred—" Burleigh's Pride," also from the Indiana Company.

At this show a clean sweep of the spoils was made by the Aberdeen-Angus. We shall first note that "Abernethy" was *second* in the two-year-old grade class, and was *first* in the like class for "cost of production;" that "Blaine" and "Logan" were *first* and *second* in yearling classes for *early maturity* and *cost of production*.

But to Col. G. W. Henry, of Angus Park, Kansas City, Mo., and his imported cow—"Bride"—belonged the honors of this memorable meeting. Having been favored with one of the colonel's herd catalogues, we cannot do better than quote the particulars as to "Bride," as there given: "Her victories at the Kansas City Fat Stock Show, in November, 1884, have never been equalled by any living cow or steer. I quote from the *Live Stock Indicator* of November 6, 1884, which tells the story of her triumphs better than I can hope to: 'All things considered, the cow "Bride 3rd" of Blairshinnoch, had the most remarkable series of successes ever achieved by any animal in one season at a fat stock show in America, if not in the world, having won premiums as follows: First, as best thoroughbred cow of any age or breed, three-years-old or over, $50; sweepstakes as best cow, grade or thoroughbred, three-years-old or over, $75; sweepstakes as best carcass of steer, spayed or barren heifer or cow, three-years-old or over, $75; and sweepstakes as best dressed carcass of any age, $100. As to her keep or preparation, her owner certifies that she ran in pasture from April to November, 1883; from November, 1883, to March, 1884, she ran in an open field with a stack of both straw and hay, but no other shelter and no other feed; from March to August 26, 1884, she ran in pasture, except during one week at the Kansas City Inter-

State Fair, and one week at the Kansas City Fat Stock Show; during 1883 she had no feed of any description but that mentioned. As he had a constant hope that this cow would breed, she had no grain, which prevented her from getting excessively fat. She having refused service from April until August, he supposed her in calf and delayed feeding her, on that account, until August 26, 1884. Since August 26, 1884, she has been kept in a stall, and consumed 5 bushels of shelled corn, 340 lbs. of corn-chop, 120 lbs. of wheat bran, 130 lbs. of oil-meal, what hay she would eat—say 750 lbs.—and nothing else. Taking into account the slight preparation given this cow to fit her for such a contest, in a show containing more than a hundred beasts, all well-nigh models, her triumph, and that of the breed, was great. She weighed, previous to killing, 1,395 lbs., and dressed 881 lbs. of net meat, or 63.15 per cent. of her gross weight. All of her premiums, amounting to $300, will be duplicated by the American Aberdeen-Angus Association, making her entire winnings in the one show $600.'"

The quality of her meat is noted elsewhere.

The late Mr. H. D. Adamson, who acted with general acceptance as a judge at this Kansas City Show, in a report he sent to the (Edinburgh) *North British Agriculturist*, thus referred to "Bride:" "Mr. G. W. Henry, of Angus Park, walked away with first ribbon for his Angus cow 'Bride 3rd' of Blairshinnoch. She is on short legs, rather square in quarter, and only half fat; however, her quality and symmetry passed her first amongst nine entries, representatives of Shorthorn, Hereford, and Holstein. A great astonishment, however, was in store for both exhibitor and the public when the little cow met the moving monster from Bow Park,

Canada—a heifer over four years of age, scaling over 2,600 lbs., or 1,200 lbs. more than herself; girth, 9 feet 2 inches.

"The Angus had symmetry only, but the Canadian, in addition to perfect symmetry, boasted of enormous weight; but her handle was most indifferent. At this time the awards were made by three practical butchers, and they took a practical view, thought the blubber useless, and gave the cow sweepstakes to the lucky 'Bride 3rd.' She was also entered for the dressed carcass prize, and was killed for that competition shortly after. Fortune did not desert her owner, for not only did the aged prize fall to her, but also the sweepstake for the best of all carcasses—a wonderful feat for a half-fed Angus, and netting $600 to her owner for his pluck. The carcass prizes were awarded by a different set of judges. 'Bride 3rd's' victory was a great feather in the cap of the Angus men, who were poorly represented in numbers, the only entry in the Angus class being a wonderful sappy yearling belonging to the Indiana Company, and showing a daily gain of 2.70 lbs., and was awarded the premium of the show for *early maturity.*"

But that was not all that the results of this show put to the credit of the "Champion Poll." The best animal "*bred and fed* by the exhibitor" was half—in this case *the better half*—an Angus. This was the cow "Burleigh's Pride," an Aberdeen-Hereford. She was awarded, in the class just noted, the *Breeder's Gazette* handsome challenge shield. This animal has created a great deal of interest. Her sire was a Ballindalloch yearling polled bull. Her dam was a yearling Hereford heifer. She was polled—a "black-whiteface," All her good points were derived from the sire. We may complete the future history of the cow here. At the show at Kansas City, the following year, she was elected best cow in the yard, weighing 1,850 lbs., still smooth and gay; the *Breeder's Gazette* characterising her as "certainly a very remarkable animal." We believe she was purchased at the late Chicago Fat Stock Show, by Dr. C. J. Alloway, late of Montreal, now of Grand Forks, Dakota, who has there started a large Angus and Hereford breeding establishment. His purpose being to show her, as a sample of the breeder's art, to his visitors.

The late H. D. Adamson, in his report, thus wrote of this cow's appearance, in her class, at Kansas City, in 1884: "Nothing, however, could touch the beautiful two-year-old black heifer cross of polled bull (Ballindalloch) on an imported Hereford cow, having 'Sir Benjamin' (1387) blood in her veins—neck, vein, shoulders, back and loins covered with the primest beef, her only fault being the light thigh, that and the white face being the only points of the Hereford. As it is always conceded that the sire has the stronger impress, Angus men naturally laid claim to their full share of the honor."

CHICAGO.

The following Aberdeens appeared: Cochrane's three-year-old "Netherwood Jock," Harvey's two-year-old "Paris Favorite," Indiana Company's "Blaine" and "Logan," and Geary's "Black Prince." There were also two grades — "Abernethy," and (Cochrane's) "Quality." J. J. Hill, of St. Paul, had also a large, fine lot of pure and cross-bred Aberdeen oxen, which had been selected by his agent, Mr. Dolby, in the north of Scotland. Unfortunately, on account of the non-arrival in time of the age certificate, they were unjustly, many thought, ruled out of competition.

"Netherwood Jock" was awarded the breed sweepstakes. "Blaine" was awarded first premium for early maturity ("gain per day"). "Abernethy" got the blue for "best carcass two-year-old and under; while "Quality" got the blue for the best dressed carcass of any animal under two.

At this show we see the last of "Black Prince." This animal had been awarded a second premium in his class at a Smithfield, London, England, and had been imported by his enterprising owner—one of the chief props of the breed. On leaving Scotland he weighed 2,600 lbs. He lost 200 lbs. in his three months journeyings, previous to his appearance at Chicago, in 1883. He was just out of quarantine, and therefore had lost *bloom*—so vitally essential to the fortunes of champion winners. "He has traveled more than any ox that ever lived." At this, 1884, show, he had one vote for the grand sweepstakes, and it was conceded that the carcass had more lean meat than any bullock of its age, size or weight, besides dressing over seventy-one per cent.

H. D. Adamson, after these 1884 shows, wrote to the *Breeder's Gazette:* "To call attention to the remarkable success the breed has achieved on the block at the late exhibitions at Kansas City and Chicago, the prizes being awarded by men who acknowledge that they have never handled the breed, dead or alive. * * * These are facts, and when the breed is numerically stronger we shall find the butchers here as willing to give the breed a special quotation as those in the old country."

It used to be an old "say" that Aberdeen-Angus

were, compared with the Shorthorn, of slower growth, during the first year of their life, but made up in the second. The average gain per day of the four Aberdeen-Angus yearlings exhibited in 1884 was 2.41; Shorthorns, 2.15; Shorthorns (1883), 1.62. Herefords are left far behind—the highest average of these being, in 1879, 2.15.

Thus, one by one, these old "says," which have been so long, without dispute, accepted for gospel, get dispelled!

1885.

KANSAS CITY.

There were of Aberdeen-Angus, the Indiana Company's "Blaine" and "Logan" (two-year-olds), Gudgell & Simpson's "Sandy" (yearling), their "Alex" (calf), and Estill & Elliot's "Felix (calf). There were of grades, Estill & Elliot's "Bloom" (two-year-old), and "Clarence" (yearling), besides "Burleigh's Pride," cow.

The breed championship went to Gudgell & Simpson's "Sandy." The *Breeder's Gazette* noting him thus: "This steer, 'Sandy,' is one of the very best blacks yet seen at these competitions, and his subsequent winning of the championship of the entire show and the medal offered by the Polled Cattle Society of Scotland (open to all ages) stamps him one of the best things yet seen from the 'doddie' ranks. He is a marvel of symmetry, well covered, and smooth as an egg from end to end. His weight at nineteen months is 1,455 lbs."

"Sandy" is a son of Gudgell & Simpson's well known prize bull "Knight of St. Patrick"—"Old Pat"—and thus a half brother to their remarkable

"Black Knight,"* declared by Mr. Wm. Watson to be the "best two-year-old he ever saw."

In all the early maturity classes of two-year-olds, yearlings and calves, the Aberdeen-Angus made a clean sweep—another proof to the contrary of the *idea* that the breed is of slow growth in its younger years. Estill & Elliot's grades took *second* places in their respective classes. In sweepstakes grade cows, " Burleigh's Pride," as already noted, was again best cow on the ground. "Sandy" was second, in the competition for grand sweepstakes, against some of the finest oxen that have yet appeared. There were no Aberdeen-Angus slaughtered.

CHICAGO.

The following animals were exhibited: J. J. Hill's " Benholm " and " Wildy," Indiana Company's " Blaine" and " Logan "—two-year-olds; Gudgell & Simpson's " Sandy " (yearling), and " Alex " (calf); of grades, Hill's " Turriff " (three-quarters Aberdeen.)

" Benholm " was awarded sweepstakes and the Polled Cattle Society's medal.

Of the class in which " Benholm " appeared, the *Breeder's Gazette* said: " The two-year-old entries comprised Mr. J. J. Hill's ' Benholm ' and ' Wildy,' and

---

*"Uncle William Watson, whose knowledge of polled cattle is probably superior to that of any man living in this country, started last week for Turlington, Neb., where he takes charge of Mr. T. W. Harvey's herd. Uncle Billy took with him his bull 'Black Knight,' now two years old, out of 'Black Cap,' and got by 'Knight of St. Patrick.' 'Black Knight' was bred by Messrs. Gudgell &·Simpson, of Independence, Mo., and cost Mr. Watson an even $2,000."—*Kansas City Live Stock Indicator*, *February 25, 1886.*

the Indiana Blooded Stock Company's 'Blaine' and 'Logan.' The best steer of the lot (easily enough), 'Benholm,' was not allowed to compete in this ring, for lack of evidence as to age, but he demonstrated his quality in good style a few days later (upon receipt of the necessary documents) by winning the consolation two-year-old championship of the entire show. The other entry from North Oaks ('Wildy') although inferior to 'Benholm' is yet a good representative of the breed."

The following is the committee's report on the prize steer "Wildy:" "The first premium was awarded a very fine model of a butcher's beast, with form, style and finish seldom excelled. Considering age this steer was remarkably well matured, and the superior handling qualities gave assurance of a carcass with a large proportion of edible meat. This steer was thickly covered with firm, mellow flesh, and with a length and thickness of loin seldom seen in two-year-old steers. The fineness of bone, nicely proportioned quarters, well filled twist, left no doubt as to the unusually large proportion of net to gross that would show to the credit of the carcass."

Of the yearling "Sandy," the *Gazette*: "Gudgell & Simpson's grand yearling, 'Sandy,' by 'Knight of St. Patrick,' stood alone in the ring for that age, and was cheerfully accorded the Society's first prize. This steer looks better every time we see him, and all through the week divided the honors with the roan Shorthorn and Mr. Earl's Hereford as one of the 'three best yearlings' in the show. He is smooth, thick, and level in his flesh, and, as a two-year-old, will be a hard one to get over. He will be looked for with considerable interest another year."

His weight was 1,470 lbs., a gain per day of 2.48, nearly 2 lbs., at one year and seven months.

The judges report of him is: "The fineness of bone, perfection of outline, gave evidence of good breeding and excellent feeding qualities, while the growth and finish of the steer left no room to doubt the skill of the feeder. The steer was a model butchers' and feeders' beast and only requires continued skillful handling to ensure a bullock of the highest degree of excellence." Such is the tribute paid by the committee. The latest accounts we have of this steer are that his progress is such as to make "rivals uneasy."

There was but one entry in the calf ring, and the first premium was awarded to the calf "Alex," exhibited by Messrs. Gudgell & Simpson, of Independence, Mo. "It would be difficult to suggest an improvement in the make-up of this calf, which only needs the same handling as in the past to ensure a very creditable and perfect specimen of the breed."—*Gazette.*

In the Consolation Class "Benholm" was best of all two-year-olds, and "Sandy" best of all yearlings in the show.

"Benholm" (weight—2 years 9 months—1,955 lbs.) drew the prize for the largest percentage of carcass to live weight—"thus repeating last year's experience" of the breed.

"Benholm" shows the highest—71.4—which is also the highest result given by any beast of any age. "To win the carcass prize in the three-year-old ring, and out-dress all competitors with this two-year-old, is certainly an honor of which Mr. Hill may well feel proud, and reflects additional lustre upon the fame of the Aberdeen doddies as prime butchers' beasts. 'Benholm' was also the winner of the Polled Cattle Society's medal, and his carcass was served by Mr. F. T. Croxson, at the Exchange restaurant, Union Stock Yards."

GRADES.

The *Gazette's* report in this section is given: " The exhibit of this year shows greater average weight, in proportion to the average age, than that of any former year. An analysis of the average weights and ages of the twenty-nine steers, shown at the late show, gives the following result: Twelve grade Herefords, shown at an average age of three years six months and five days, give an average weight of 1,070 lbs.; an average gain per day, from birth, of between 1.60 and 1.61 lbs. Thirteen grade Shorthorns, at an average age of three years six months and sixteen days, show an average weight of 2,037 lbs.; an average gain per day, from birth, of between 1.57 and 1.58. This showing is rather favorable to the get of the Hereford bull; but the three crossbred Angus-Shorthorns, shown by Mr. J. J. Hill, outdo either Shorthorn or Hereford grades. * * * The trio of this blood, shown from North Oaks, at an average age of three years four months and nineteen days, gave an average weight of a fraction over 2,271 lbs.; an average gain per day, since birth, of about 1.88 lbs.

"The thirteen grade Shorthorn yearlings shown had an average weight of 1,257 lbs., at an average age of nineteen months and nineteen days, an average gain per day, since birth, of 2.15 lbs. The sixteen grade Herefords came into the ring with an average weight of 1,375 lbs., at an average age of twenty-one months, average gain per day of nearly 2.21 lbs. The Aberdeen-Angus entry, Mr. Estill's 'Flash' weighed 1,360 lbs. at twenty months, an average gain per day of 2.25 lbs.

Mr. Lucien Scott's 'Last Chance' (grade Holstein) weighed 1,300 lbs. at twenty-two months, average gain per day, since birth, 1.89. The average age of the thirty-one animals shown was twenty months and fourteen days; average weight 1,322 lbs. (within a fraction); average pain per day 2.17.

"Mr. J. J. Hill's polled Scot 'Turriff' (cross-bred Angus-Shorthorn), yielding 1,404 lbs. (dressed weight), and netting 68.1 per cent., had the honor of drawing the prize for *quality* of (*most edible*) beef. Mr. P. D. Armour, of Chicago, purchased the carcasses of 'Hutcheon' and 'Kinloss' (and we think of 'Turriff' also), distributing the beef among his friends."

The *Farmer's Review*, Chicago, in an editorial article, November 25, 1885, said: "Edible meat is or should be the standard aim of the breeder and exhibitor of fat stock"—so that the point of the following comment from the same paper will be apparent: "Breeders of Scotch hornless cattle were not a little consoled when J. J. Hill's grade Angus steer, 'Turriff' carried away the premium for best three-year-old carcass and also prize for carcass showing most edible beef. Having gained these two most important awards it was natural to suppose that sweepstakes would also have fallen to his lot. But no; for a grade Hereford, called 'Joe,' bred by Seabury & Sample, took this."

The *Gazette* said the "exhibit was small, but first-class." The *National Live Stock Journal* said it was "small, but select." We are satisfied with the record not only of this, but of the previous two years. We are satisfied that it only requires a short time for the breed to "Clement-Stephenson" all others on this side, as it has done so often in Europe—and to make our rivals permanently "uneasy!"

JUSTICE (1462. See p. 109.

CHAPTER XI.

APOTHEOSIS.

PART. I.

THE CENTENARY SHOW OF THE HIGHLAND SOCIETY.

It is not intended to go into a history of "the greatest victories ever achieved by cattle—which have been won by the Aberdeen Polled," in France and Bvitain. That would be writing the chief portion of a separate undertaking, which would require to be entitled "History of the Aberdeen-Angus Breed in the Show-yards of the World."

We shall just enumerate the most colossal of these, by which, in the march to fame, they have secured supremacy. The first great demonstration they made —proving to the world that they were a breed of the first order—was at Paris, in 1856, thirty years ago, when they were awarded "the great gold medal of France" for their superior excellences.

In 1862 they were, at Poissy, France, awarded, "in the person" of a noble steer, the championship of the world, and had bestowed on them the Prince Albert Cup, value $500.

In 1878, again at Paris, their *double victory* need hardly be alluded to.

Coming to England, at Birmingham they have been champions in 1867, 1871, 1872, 1883, 1884, 1885. At Smithfield they were champions in 1867, 1871, 1881, 1885, The *double* victories of the years 1881 and 1885 are specially and peculiarly rememberable in the annals of show-yard stock.

These do not take into account the instances in which they have been *best female*, or *best ox*, or *reserve ;* or again, the cases in which they have supplied *half* the champions by crosses. These are the greatest victories ever achieved by animals of the bovine kind in open competition with the world.

For several reasons it is desirable to give to the world a compiled synopsis of deeds and opinions of the 1884 Centenary Show of the National Scottish Society; also of the year 1885, at Birmingham and Smithfield. The latter are—though but "one" of the greatest victories—the most recent. And by giving this analysis and selection, it will serve to show to the breeder on this side what such a record of victory means and establishes. It is of surpassing interest to others besides the enthusiasts of the champion poll.

These, representing the breeding and feeding demonstrations of the breed, will ever be of prime importance to Aberdeen-Angus breeders, so we first give an account of

THE GREAT CENTENARY MEETING OF THE HIGHLAND SOCIETY.

We extract from the official report of the Centenary Show of the Highland Agricultural Society, at Edinburgh, 1884, written by Rev. John Gillespie, M.A., of Mouswald, director of the society, and editor of the "Galloway Herd Book," the following on the "Polled Angus or Aberdeen:" "The entries of this splendid and deservedly popular breed of cattle were the most numerous in the show-yard, there being 53 bulls and 104 cows and heifers, making the very handsome total of 157 head. At no former show of this or any other society has there been such an extensive display of the northern polls, and the quality of the exhibits was even more conspicuous and remarkable than the number. Indeed, it may be safely said that there was no department of this Centenary Show, either animate or inanimate, which made so deep an impression on the vast crowds of visitors as the Polled Angus or Aberdeen cattle. Every class contained a large number of animals of the highest personal merit, and they were brought out in a state of perfect bloom, which reflected the greatest credit on the exhibitors and herdsmen who tended them. It not unfrequently happens with the turn out of all breeds, at shows, that a numerous entry leads to a lowering of the average merit, but far from this being the case here, the average excellence was never so high in most of the classes, notably in the aged bulls and cows. The judges had a difficult task in awarding the prizes where the competition was so wide and the merit so evenly balanced, but they discharged their duties with the utmost care and discrimination.

"The fifteen aged bulls in the catalogue comprised no fewer than three first prize winners at former shows of the society. These were Sir George Macpherson Grant's six-year-old Justice (1462); Mr. Anderson of Daugh's seven-year-old Prince Albert of Baads (1336); and Mr. George Wilken's of Waterside of Forbes' four-year-old bull The Black Knight (1809). The greater part of the time occupied with this class seemed to be spent in determining whether Justice (1462), or Prince Albert (1336) was to be awarded the much coveted ticket. They are both magnificent bulls, and increasing age has in no degree impaired their splendid qualities. The fiat was eventually given in favor of Justice, which is an extremely gay, attractive bull, showing beautiful symmetry and quality, with an immense quantity of flesh on the most valuable parts. Prince Albert has never been beaten in a lengthened show-yard career against the best of his day, and he was a popular favorite, as he was well entitled to be, with his immense level frame covered with a thick covering of flesh, and wonderful activity for his age. Justice having won in this hotly contested fight, it was a foregone conclusion that he should win the cup as the best male. The Black Knight, which was placed third, was not in quite such good form as at Inverness in 1883, when he was first in the aged class, but he is deep and thick, and altogether handsome."

"The two dozen cows forward out of the 34 entered in the catalogue, presented one of the most magnificent spectacles, not only in the breed, but in the show-yard itself. The dozen chosen in the short leet were splendid cows. The task of the judges in this competition was an extremely arduous one, and they took every pains in the discharge of it. Eventually Mr. George Wilken, Waterside of Forbes, was assigned the coveted first ticket, with Waterside Matilda 2nd (6312), a three-year-old cow, first in the two-year-old class at the show of the Highland Society at Inverness, in 1883, and which was the champion at Aberdeen the week preceding the Centenary Show. She is altogether an almost model cow, being round, straight, thick, and level. She was followed closely by Sir George Macpherson Grant's Electra (4186); the five-year-old Erica cow from Ballindalloch, which was placed second. She has true feminine character, fine in bone, and specially good in her forequarters, though she looks a little deficient in her hocks."

Justice took the Cup as the' best male, and also headed the Ballindalloch first prize group.

Waterside Matilda, the winner in an unrivalled class of cows, headed Mr. Wilken's first prize family.

"THE INCOMPARABLE."

Justice has also won The McCombie Prize at Aberdeen. He is out of the original Jilt (973) by Elcho; which was out of the original Erica (843), by Juryman (404), the first Jilt bull bred at Ballindalloch—

$$\text{Justice (1462).} \begin{cases} \text{Elcho (595).} \begin{cases} \text{Juryman (404)).} \begin{cases} \text{Bright (434).} \begin{cases} \text{Black Prince (366.)} \\ \text{Nourmahall (726.)} \end{cases} \\ \text{JILT (973.)} \end{cases} \\ \text{ERICA (843.)} \end{cases} \\ \text{Jilt (973).} \begin{cases} \text{Black Prince (366.)} \\ \text{Beauty of Tillyfour 2nd [1180.)} \end{cases} \end{cases}$$

Justice is thus a grand living example of the happy effects of the "inter-nicking" of the Ballindalloch "twin-star" tribes—the Ericas and Jilts.

In a spirit worthy the title of *national enterprise*, this great modern instance of show-yard success and supremacy, the modern Champion and Representative of the Breed—a perfect King of Beasts—has been imported for use in The Goodwin Park Herd and The McCombie Herd, personally by that well known doddie enthusiast, Judge J. S. Goodwin, of Beloit,

Kansas, where in the future he may be seen and "heard from."

As pointed out by the *Breeder's Gazette*, of Chicago, and the *Live Stock Journal*, of London, England, there is a peculiar interest in this transaction—to consummate which Mr. Goodwin made a special trip to Britain —insomuch as it was in The Goodwin Park Herd that great Judge, the very distinguished brother of Justice, had his last abiding place. Breeders of polled cattle on this side will hail the advent of Justice with enthusiasm, and will, as one has expressed it, *throw up their hats over it!*

Prince Albert, through his Ballindalloch sire Bachelor, traces to much of the same Jilt blood—as, to Juryman (404), and Trojan (402) by Black Prince (366.)

The Black Knight shows the same on both sides; sire of dam of sire is Black Prince (366), sire of dam is Black Prince (366), and sire of grand-dam is Black Prince (366.)

Portraits of Justice may be found besides the one given here, which was engraved by the publishers of *The Live Stock Journal*, London, in Macdonald & Sinclair's "History," and in Vol. X. of "The Polled Herd Book;" of The Black Knight, Waterside Matilda 2nd, and Flush 2nd, also in Vol. X. of "The Polled Herd Book."

Of the cows, Waterside Matilda 2nd (6312) is by Pride Knight of the Shire (1699), and Erica Electra by Pride Petrarch (1258)—the former bull a McCombie Cup winner; the latter a Paris, 1878, winner. Flush 2nd has Black Prince, Windsor, and Napoleon blood mostly.

So that at this memorable show we have the Erica, Jilt, Queen and Pride blood entirely producing the

THE BREED THAT BEATS THE RECORD.

winners, No better test could be stated of the potency of "blood."

The Live Stock Journal editorially remarked on this show:

The exceptionally fine display of Polled Aberdeen-Angus cattle has been the subject of general comment during the show. This breed has made rapid strides during the past ten years, and no one who has seen the polled collection at Edinburgh, this week, would dispute its claim to the possession of very great excellence. These cattle show no lack of size, and while most of them have been highly fed, their levelness of flesh, fineness of quality, and freshness in gait, are quite remarkable. The polled cow class is, in regard to average merit and numbers, one of the best classes of live stock we have ever seen in any British show-yard. The leet of ten or twelve which were drawn out by the judges, for the final tussle, formed an array of animal perfection such as we have rarely seen equalled.

Mr. Geo. Hendry, of the *Daily Free Press*, Aberdeen, one of the few good live stock reporters in Britain, thus writes to an American journal:

In Scotland, as elsewhere, we have the claims of one breed canvassed and pitted against those of another, and a slight spirit of jealousy existing among our breeders, so that all eyes were eagerly turned towards the Centenary Exhibition of the Highland Society, which it was felt would be a good test of the merits of our world renowned varieties of cattle. Briefly and frankly, then, none stood that test so well as the Polled Aberdeen-Angus, which made a surpassingly grand appearance in the show-yard, as rival breeders were forced to admit. Had this been the first time they were known to fame, their reputation as the breed, *par excellence*, most calculated to give the greatest return to the feeder, would have at once been made, but they have a well established character in this respect, which will be strengthened and extended by the event of last week. * * * The one hundred and fifty odd entries of Polled Aberdeens undoubtedly carried the palm in the cattle sections. How wonderfully well they carry their flesh all over their round, sleek bodies, and that, too, though many of them have been somewhat injudiciously forced for the show-yard! How nicely set on their legs are they, and how clean made and small in the bone! *Beau ideal* specimens, in short, of cattle that are famed throughout the world for their quality and wealth of flesh. Never, on any former occasion, has such a magnificent display been witnessed, for not only were the winners of great individual merit, but the quality was well sustained in the very end of the different classes.

PART II.

BIRMINGHAM AND SMITHFIELD IN 1885.

A Scottish correspondent of the *Breeder's Gazette* wrote, December 31, 1885: "Never before has it fallen to the lot of any breed of cattle to make such a marked sensation among all classes of breeders and feeders as it has fallen to the Aberdeen-Angus to do this year. Not only has the championship of the two great shows at Birmingham and London been awarded to one of the breed, but the general excellence of the exhibits has been universally admitted; and it is not only in the classes for pure-bred cattle that the Aberdeen-Angus have made their mark, but in the class for cross-breds the blacks have had it all their own way. The judges at Smithfield, only one of whom, by the way, was a Scotsman, seemed unable to recognize anything but what was more or less impregnated with the prepotent blood of the 'black, but comely.' * * * Be it remembered that the word 'polled' as used here, may be taken as strictly applicable to the Aberdeen-Angus breed. At the great shows this year, as in former years, Galloways have been remarkable from their absence; in fact, I trace Galloway blood in only one of those appearing in the prize list, viz., the second prize steer in the old cross-bred class, which is out of a Galloway cow."

THE BREED THAT BEATS THE RECORD. 113

"Delta," the special London cerrespondent of the *Gazette*, said of the Birmingham show: "The best butcher's animals at the exhibition, were the famous beef-breed—the Polled Aberdeen. These animals made a good show, and were animals of style and character." "Delta" made similar remarks on the Smithfield show.

"Sigma," correspondent of the *National Live Stock Journal*, writes thus of the polls at the Birmingham show: "The chief honors of the show have fallen to the Aberdeen-Angus, this breed supplying the champion, and having half the credit of producing the reserve, as well as a good many of the best animals among the first rate section of cross-breds.

"One of the best classes in the show was that of Scotch Polled cows and heifers. only Aberdeen-Angus, however, putting in an appearance."

Particularising Luxury, he proceeds: "This beautiful specimen of the breed is perfect in symmetry, even in flesh, small in size, very true in character, and with fine bone. Mr. Stephenson, her breeder and owner, has now achieved the unparalleled feat of winning the Birmingham championship three years in succession. He is the county veterinarian, and is well known as a successful breeder and feeder. [He was formerly a hot Shorthorn fancier]. His accession to the ranks of polled cattle breeders is a great strength to that body, and Scotchmen need his aid at present, for they now greatly miss from among them an enterprising pioneer like the late Mr. McCombie, of Tillyfour."

Of Smithfield, the same authority writes: "The Smithfield Fat Stock Show has resulted in a brilliant series of victories for the Aberdeen-Angus breed, and its champion Mr. Clement Stephenson, Newcastle-on-Tyne. Mr. Stephenson's heifer, Luxury, which carried

LUXURY.*

off the blue ribbon at Birmingham, was soon singled out here as the most likely winner of the best prizes, and this expectation was fulfilled. * * * Her most dangerous antagonist, too, was another Aberdeen-Angus heifer, exhibited in the same class. This was Sir William Gordon Cumming's two-year-and-nine-months-old heifer, got by the Queen bull Dustman, and from a dam by the Pride bull Black Watch, a very thick, plump animal, and splendidly ripened. The class in which these two animals competed was the best in the hall.

---

*Acknowledgment is here extended for the courtesy of the proprietors of the *Breeder's Gazette*, Chicago, Ill., for permission to use the above excellently executed cut.

"The strength of the Scotch Polled breeds—Galloways being again 'conspicuous by their absence'—was in the female classes. Than the half score Aberdeen-Angus cows and heifers, I may repeat, nothing more choice has ever been witnessed at Smithfield.

"The cross-breds, another grand section, formed a striking testimony to the excellence of the Aberdeen-Angus and Shorthorn cross."

The *Free Press*, Aberdeen, Scotland, alluding to Mr. Stephenson's victories for the last three years, at Birmingham, said: "These victories of Mr. Stephenson have placed the polled Aberdeen-Angus breed on the highest pinnacle of fame as beef producers. We are sure no one—a Scotchman least of all—will grudge Mr. Stephenson the honors which he has won so worthily."

Of the Smithfield show it said: "Scotch feeders and breeders are jubilant to-day because of the great achievements of the Polled Aberdeen-Angus cattle at the eighty-eighth annual show of the Smithfield Cattle Club. This breed was never seen to better advantage in London, or for that matter anywhere else, and the concluding victory of Mr. Stephenson's polled heifer Luxury over the best of all breeds was a cause of much rejoicing among the Scotchmen in the hall. None can deny that this year the Polled Aberdeens have been the very cream of the great London exhibition. There were at least a round dozen of black beauties that would have with the utmost ease put to rout a like number drawn from any other breed in the exhibition. The year 1885, at Smithfield, may truly be called the year of the Polled Aberdeens. The show has been a remarkable one in several ways. In the first place it is all round one of the best exhibitions probably that have ever taken place in London—its merit consisting in the general high character of the stock rather than in notable ' plums.' The average merit is much better, and that, after all, is the chief consideration at an exhibition of this kind. Last year's show was, from a utilitarian point of view, ranked high, and the present one might be placed even higher in that respect. A notable feature is the magnificent array of cattle that in crossing had been clearly indebted to the blood of polled stock for their outward moulding.

" There was no mistaking the significance of the keen interest with which the judging of the Polled Aberdeen or Angus cattle was watched. The competition, by the way, was open to Galloways, but not a single specimen of that breed was shown. There were 25 entries in all, as against 20 in 1884, and 17 in 1883, and such a grand display never before graced any show-yard. If anything is calculated to revive the demand for polled cattle, it is the exceptional merit of those shown by Scotch and English exhibitors to-day. As formerly noticed, a new class has been added for young steers."

It described the class of polled cows " as not excelled by any in the show." And in fully describing Luxury, after all her journeyings, happily hit her off by saying she " looked *as fresh as paint.*"

The *Banffshire Journal* says of the London show: " The class in which Luxury appeared was undoubtedly the best in the show, and the best of the breed that has ever been together."

The *North British Agriculturist*, December 2, 1885, in its report of the Birmingham show, says: " The champion heifer is one of the best-matured and neatest animals at her age it has ever been our privilege to see, and presents a beautiful outline, fine bone, and attractive gait. * * * It is worthy of remark that this is the third successive time in which Mr. Stephenson has carried the Birmingham championship with animals of the Polled Angus breed, and that he has thus proved himself one of the most successful breeders who have sought to vindicate the beefing properties of the favorite breed."

In its report of the Smithfield show the same paper says: "If there has been one section of the great Smithfield show, which is now being held in the Agricultural Hall, Islington, London, more remarkable for its recent development than another, it is undoubtedly that of Scotch polled cattle. Never, during the long and eventful history of the exhibition, have the favorite 'black-skins' figured so prominently among the beef-making breeds of British cattle as on the present occasion. The Aberdeen-Angus breed not only bears the distinction of producing a worthy champion, but it may safely be regarded as the feature of the show. It was predicted by some vigilant breeders last year that the polled breed would this year be further to the front than ever, and their predictions have not been falsified. On the other hand, the most hopeful anticipations have been fully realized, and the highest ambition of the promoters of the breed fully gratified. Be this as it may, it has been observed by visitors to the exhibition at Smithfield, for some time past, that, year by year, 'black-skins' have been becoming more numerous. *For beef,* black is the dominant color; and even Shorthorn fanciers, on Monday, in giving the awards, evinced a predilection for the sable coat.*

---

*See pp. 18, 19.

"The exhibition all over is one of the best ever held at Islington. Indeed, some people would put it down as the finest show of the kind ever seen in the Agricultural Hall. As regards size, the exhibition exceeds that of last year by eight entries, and is considered larger than the great majority of its predecessors. It comprises a total entry of 543, of which there are 293 cattle, 190 sheep, and 60 pigs. Out of this number there are only some twenty-four absentees, and many of the classes, notably those of Scotch cattle, are bigger than they have been in recent years. Mr. Stephenson is again the champion with the handsome little heifer with which he accomplished the usual feat at Birmingham last week—where he won the championship for the third successive time with representatives of the polled breed. His greatest success in the feeding of polled cattle is something to be proud of."

It describes as follows the chief features of the show:

"ABERDEEN-ANGUS CATTLE.

"No apology need be offered for departing from the order of the catalogue, and referring in the first place to the Aberdeen-Angus cattle. We need scarcely repeat that they form the feature of the show, and that they are more numerously represented than on any former occasion.

"Polled heifers formed a magnificent class of nine. The contest here was keen and protracted, and, as was to be expected, resulted in a victory for Mr. Stephenson's famous Luxury. Some of the judges seemed to like the Altyre heifer fully better than Luxury, but good though the former is, the little champion could not be got over. The Altyre heifers, both of which are very nicely brought out, are fully as attractive at first sight as Luxury, but the heavier of the two, which weighs fully 15½ cwt., is nearly 1½ cwt. lighter than Mr. Stephenson's heifer. The second heifer, sired by Dustman, is a good, massive animal, but not so evenly balanced as Luxury, while she is more defective than her successful rival in the shoulders. She was second at the Aberdeen Highland Show last summer. The Haughton heifer, which was second at Birmingham, occupies the third place, and is followed by another daughter of Dustman from Altyre. Favourite of Altyre, is well covered, but she is a trifle soft in her flesh. She was second yearling at the Edinburgh Centenary Show of the Highland and Agricultural Society."

It is conceded by the English papers that the Polled Aberdeen exhibits constituted the feature of the shows, and the premium lists prove that they bore off the lion's share of the honors.

The *Field*, London, in describing the London show, says: "The strength of the Scotch cattle, and the most meritorious element of the bovine exhibition was found in the polled cattle, which was one of the largest and best collections ever seen in London."

The *Live Stock Journal* is the leading national journal of its class and a model in every way. In its report it holds the balance evenly. After adverting to the "cry" that has been, since 1880, for early maturity, it remarks that, with the single exception of the Norfolk class, all the breed cups were won by animals under three; and after giving full justice to Mr. Stephenson as having done the never-dreamed-of-as-possible feat of winning, against all Britain, the Elkington Challenge Plate three times in succession, and with Polled Aberdeens, says: "The Polled Aberdeens have accomplished a feat unparalleled in fat stock shows." Of the Smithfield achievements words failed this journal—contenting itself by saying: "The Aberdeens have surpassed even all their former appearances."

"The victory on this occasion," says the *Free Press*, "the only time that the championship at Birmingham and Smithfield has been won in the same year and by the same animal," is further enhanced, as the *Live Stock Journal* rightly points out, by "the animal being also bred by the exhibitor," thus winning, at Birmingham, the extra Prince of Wales' (or president's) cup, and, at London, the very coveted great gold medal of the show; besides the class prizes—cups for best of breed, best of sex, best of all—at all these places.

The *Agricultural Gazette* contains a report very reminiscent of the old hands. It says: "That the breeders of this excellent variety for beef are determined not to lose any opportunity of pushing their cattle may be seen by the names of the herds which are represented. One was bred at Ballindalloch Castle; others came from hardly less noted breeders; and it is long since Warlaby or Wetherby furnished a specimen ox for graziers to pattern against other varieties. There can be no doubt that it has been the wisest policy which has made first Sir W. Gordon-Cumming, and later on Mr. C. Stephenson and Mr. G. Wilken, devote to the show-yard heifers and steers which breeders, more greedy of immediate returns, would have sent to the summer shows if they exhibited them at all. Probably, no one breed had such high types of its best tribes, 'fed for slaughter,' as had the black polls. And, as a consequence, by common consent, this is 'a black year' at Islington. The pure bred polls (and those produced by using polled bulls to Shorthorn cows) being very prominent."

Next to her stood, in her class, no less an antagonist than Sir W. C. Gordon-Cumming and Robt. Walker's representative. This heifer was by the Queen Duchess bull Dustman (1667), the sire also of the Altyre second prize ox in the young steer class. Some nice points come into consideration now. By the reports Altyre's star would, on this occasion, seem not to have been on the ascendant. The reports even indicate that there were onlookers who would have put him first in both classes of heifers and young steers—even before Luxury! Certainly their exhibits were not much behind their celebrated two of 1881, and the *Agricultural Gazette* says: "It is right to say that, standing by Sir W. C. Gordon-Cumming's pair, she (Luxury) by no means made these look rough. They pressed her hard. Both of these heifers were bred by the exhibitor, and were by his Dustman (1667), a bull whose progeny have remarkable hair—as, see the second prize steer between two and three years old. This steer had a half brother which was second in the younger class (to George Wilken), and even of greater promise. * * * Although Mr. C. Stephenson's herd won the great prize his herd certainly has not surpassed Sir W. C. Gordon-Cumming's, all of whose entries are of one pattern, and that a very fine pattern. The comparison of the entries in those classes show that this variety is becoming more and more uniform every year." These remarks are but just to the excellence of the Altyre cattle and no disparagement to Mr. Stephenson's. It is certain that if Luxury had not been present or had been beaten by Altyre, that Altyre would have gained the championship. So that here we have, as it were, a waste of one champion altogether. And further to quote a correspondent of the *Live-Stock Journal* (London): " The question arises whether the second-prize polled heifer beaten in her class by the champion was not the second-best beast in the show. Any way she should have had a chance, and should have been placed in competition for it." This is one of the certain disadvantages of the working of the London system.

Speaking of the class of yearling Angus steers the same paper says: " Indeed the whole seven, shown in this class, average 12 cwt. 2 qrs. 2 lbs., though not exceeding 20 months in age. This must be held an enormous advance in early maturity with what southerners call ' Scots.' " And of the weights of the class of cows: " These weights would have been incredible to a Norfolk grazier of this variety forty years ago."

How often have we been present in that great crowd of Englishmen during the intense excitement of the awarding the championships! " The last great award, the championship, has yet to be decided. Excitement runs high. Royalty joins in the pean of admiration. A ringing cheer echoes through the great hall when at last the 'blue ribbon' is placed on Mr. Clement Stephenson's Aberdeen-Angus heifer, Luxury, proclaiming her 'champion,' the bucolic Queen of the day." The crowd breaks through the charmed line kept by the now overcome policemen erstwhile of supremest dignity. The champion is daintily handled from poll to tail. Even Royal gloved fingers deign to come in contact with the model form. Congratulations pour over the lucky owner from prince, peer, people. Hats are thrown up, cheers break through the magic silence, the northern Doric is loosened over assembled Cockneydom. Once more the blue is "over the Border," once more the victory is to *Heather bloom.*

The *Banffshire Journal* reporter writes thus as to the interest taken in Luxury: "It is curious to notice how popular attention centres upon the leading prize winners. Mr. Clement Stephenson's champion Polled Aberdeen-Angus heifer, Luxury, was continuously surrounded by a crowd, and would never have been allowed to rest had the populace had their own way. Happily, however, Mr. Hine, the secretary of the Club, and the Stewards, have a proper appreciation of the treatment of cattle. Luxury was penned up along with a neighbor, and when the steward was absent, a policeman protected her from the hands of the curious. She is of an admirable sort, having gone on feeding during all the show time, so that her fine appearance is fully maintained, and will be reflected in the photographs of her that have been secured. Three different photographs were taken of Luxury, and I learned that one of them was intended for the Polled Cattle Society. It was understood that at one time Mr. Stephenson intended to take Luxury back to his farm, near Newcastle-on-Tyne, and try the experiment of reducing her condition and breeding from her. No sooner, however, had the champion prize been awarded than Mr. Stephenson was besieged with offers for the heifer and she was sold at, we believe, the net price of £150, to Mr. Grant, butcher, King William's street, Charing Cross, London. Mr. Grant is, in the ordinary course of his business, an extensive buyer of Aberdeenshire beef, and knows full well the public appreciation of the first-class article."

Luxury won £364 in prizes and was sold for £150 for beef at the rate of 2s. 6d. per lb. The following are particulars of her test at the block. The *Banffshire Journal* of January 5, 1886, makes the following remarks: " In reporting on the recent Smithfield Club Fat Cattle Show, we direct attention to the decreasing number of aged stock and the growing favor for fat cattle at two years old and under. The substitution of a dead meat for a live cattle trade among Metropolitan butchers accounts to some extent for the increasing demand for prime young cattle. When an ox above three years old is cut up, there is far greater proportions of low priced meat than in a two year old. The inferior parts do not sell so rapidly nor at relatively such high prices as was common a year or two ago; while the taste for the roasting parts has increased and its value is well maintained. The great characteristic of the Polled Aberdeen-Angus heifer Luxury, with which Mr. Clement Stephenson won the champion prizes both at Smithfield and Birmingham, was the apparent large proportion of the finer pieces of meat that she carried and the small appearance of bone and offal. Luxury was slaughtered and made into meat for Christmas. The carcass was inspected by the representative of the *Mark Lane Express* who has been a consistent and able exponent of the view of a ' killing test ' of fat cattle. Mr. Grant, butcher, Charing Cross, London, had Luxury killed. Her age was 974 days, her live weight 1,724 lbs., and her average gain in weight per day for that period was equal to 1.77 lbs. The carcass when quartered appeared to have no coarse meat at all; there was no more scrag than in a sheep, the neck being a little thin flap coming out of the immensely thick shoulders, whilst the shins were exceedingly small. The smallness of the bone in proportion to the thickness and weight of the carcass was something very much out of the common way, and the proportions of the animal could be seen to very much more advantage when the carcass was hung up than when it was alive. The salting meat, bottom of ribs and flank, was very fat, but all along the back, from head to tail, there was a deep covering of lean meat of the finest quality marbled to perfection and not beyond it, with a shallower covering of outside fat than many of the show animals. Taking the fat in proportion to the lean, the quality of the lean meat, the large proportion of meat to bone, and the very large proportion of prime pieces to the inferior cuts, this animal would have come first had there been a dressed carcass competition. In America, this year, the 'best killing' animal of any breed was, as we recently noticed, a Polled Aberdeen Angus two-year-old ox. Among show animals the breed may therefore be held to have taken the first place in regard to early maturity. The fact that the Aberdeen-Angus cattle stand at the head of the Metropolitan live cattle market may be taken as an indication that the qualities appearing in the specially fed animals are characteristic of the whole breed. The movement for a slaughter test in connection with our leading fat stock shows is very likely to be fully discussed before our next Christmas exhibition."

The London *Live Stock Journal* in an editorial paragraph says: " In our notice of Mr. Clement Stephenson's Aberdeen-Angus heifer, Luxury, the champion at the Birmingham and London fat-stock shows, it was remarked that we never remembered having seen a greater weight of choice meat carried in such a small compass. The appearance of the carcass and the weight amply bear out this opinion. Her dressed carcass weighed no less than 164 stone, 6 lbs., and her's was 'the most perfect body of beef that was ever seen, and not at all fat.' The live weight of Luxury was 15 cwt. 1 qr. 6 lbs., or 1,714 lbs., and her dressed weight was 1,318 lbs., showing as much as 76 per cent. of dead meat to live weight. It may be doubted, if this has ever been beaten by any well-authenticated record."

The above statement was received with incredulity in America, but the following from the editorial columns of the same journal for March 12, 1886, puts the matter at rest: "The statement made in these columns as to the carcase weight of Mr. Clement Stephenson's champion Polled heifer Luxury has not unnaturally excited a great deal of attention, and it has been suggested that the very high percentage of 76 may have been due to miscalculation. Mr. Grant, 5 King William street, Strand, London, W. C., the purchaser of Luxury, and upon whose figures our statement was based, has written the following letter to Mr. Stephenson, which will remove any doubt as to the authenticity of the return:—'In reply to your letter, I beg to say the weight sent you previously of Luxury was quite correct. I can quite understand any one being deceived in the weight, for I must say she was the most solid body of flesh that ever I saw or bought. The weight given you was when all fat was removed.'"

Luxury (7783) belongs to a Portlethen family of "ancient lineage" originating from the old Ardovie foundation of Lively (256) by Earl o' Buchan (57). For a complete history of "How Luxury was bred" see the *Breeder's Gazette*, February 18, 1886.

CHAPTER XII.

APOTHEOSIS—Continued.

PART III.

THE ABERDEEN CROSSES.

There is one great series of facts not yet noticed that are even more calculated to revive and enhance the interest in the Polls. We of course allude to the prepotence of the Aberdeen sire as displayed by the record of the classes for crosses at the English exhibition. Breeders in England have been gradually but surely educated in this matter, till what is the result to-day? the supremacy of the Aberdeen in this, as in all other points, proving the Aberdeen sire is prepotence, superior always to the Short-horn, Hereford, and all the other breeds. The *Free Press* remarks again of the class at Birmingham: "Than the cross-breeds, there was no better feature of the show." At Smithfield: Again, (same report): "A notable feature is the magnificent array of cattle that in crossing have been clearly indebted to the Polled blood for their outward moulding."

"Of cross-bred cattle there was an extra fine display, the entries showing a large increase on former year, and the average quality all around was of a high character. Reference has already been made to the important part the polled breed has played in the moulding of many of the crosses which have been exhibited in England this year, and one result of the present campaign will probably be a large increase in the number of polled cross-bred cattle in England. * * * The animals in running for the Cross cup were all typical polls, a fact to which attention was prominently called in the hall. * * * A great feature of the exhibition, it is understood, will be the grand display of Scotch-Polled pure-bred and cross-bred cattle. The Polled Aberdeenshire which have come into England are having a great influence upon the southern breeds, and wherever they come into contact, the northern poll conquers, and leaves his own type indissolubly fixed upon the animals which inherit his blood. If the process goes on with the same speed as it has begun, one will soon find all the cross-bred cattle at these exhibitions Black and Polled."

The *North British Agriculturist* of December 9th, 1885, writing of the same show, says: "In this department many fine animals are shown. There is a preponderance of 'black-skins,' and almost all the principal winners are closely related to the Aberdeen-Angus breed. Strange to say, the judges, though of Shorthorn leanings, turned aside with wonderful dispatch all the roan animals, and wrought strongly for the interests of the Polled breed."

The *Mark Lane Express* contains some rather striking comments suggested by the cattle exhibited at Birmingham. Among other things it says: "The standard of the show, so far as fat stock are concerned, was, on the whole, quite equal to that of former exhibitions, yet it differed very essentially from many recent shows in one very important and significant feature, which was this—the weakness of the classes for pure breeds was more than counterbalanced by the extraordinary strength and merit of the cross-bred section, which was, roughly speaking, fully one-half Scotch blood. It must have been patent to every careful observer that the Scotch breeds and the Scotch crosses, the latter more especially, made the Birmingham show of 1885 what it was Of course, the crosses could not be produced were it not for the maintenance of pure-bred herds of the several breeds, but it is plain that the day has gone by for fancy values to be attached to long pedigrees. The want of the day is good animals of good breeding, eligible for entry in herd books with the oldest fancy strains, which may be depended upon to get useful stock for hard-working tenant farmers' use."

The *Field* speaking of crosses at Smithfield says: "Cross-bred cattle were a marked feature of the show, both numerically and for individul merit which was great—the Scotch Polled Aberdeen-Angus breed having a marked influence on form and color. It may be taken that with few exceptions, notably Mr. Wortley's champion steer of 1884 (Hereford-Short-horn) a cross between Polled Scot Aberdeen-Angus and Short-horn in some degree or other, is the most successful, both as to form, quality and early maturity."

The other journals have similar remarks, the *Live-Stock Journal* saying it was remarkable that in "nearly every case the judges preferred for the chief prizes, animals of the Polled type."

CHARACTER OF THE PRIZE WINNERS IN THE CROSS-BRED CLASSES.

The following summary has been prepared from the details given in the above-noted journals and will show the composition and character of the classes. It may be useful and interesting:

AT BIRMINGHAM—THREE-YEAR-OLD CLASS.

1st. Sire Short-horn Grand Duke of Oxford 32953; dam Polled Aberdeen Queen Mary of Glamis 3312. This animal also took the Cross cup at this show, but got no further, though he had been champion at Norwich over Mr. C. Stephenson's Luxury, where the reporters declared their preference for the latter. This ox, put back at this show, was, as we shall see, put back still further at Smithfield. Mr. Wm. Watson makes the following comparison: "Mr. Loder's ox weighed 2,624 lbs., a younger steer than Chicago's champion Regulus, yet weighing 334 lbs. more, showing the decided superiority of the Aberdeen-Angus and Short-horn cross to that of the Hereford and Short-horn." Character—Black Polled.

2d. Sire Polled Aberdeen; dam Short-horn; color blue-gray. Highly commended (two)—details not sufficient.

TWO-YEAR-OLD CLASS.

1st. Sire Polled Aberdeen; dam Short-horn; Black Polled.

2d. Sire and dam not stated in reports; red-and-white.

3d. Sire and dam not stated in reports; roan. Highly commended, blue-gray; highly commended, roan.

THE BREED THAT BEATS THE RECORD. 131

YEARLING CLASS.

1st. Sire Aberdeen Proud Knight 1922; dam Short-horn; Black Polled.
2d. No details.
3d. Blue Poll.
4th. Polled and roan.

COWS AND HEIFERS.

1st. Sire Polled Aberdeen; dam first-cross (Aberdeen-Short-horn) Black Polled., The *Live-Stock Journal* said of this heifer while at Birmingham: "It may be doubted if any animal in the show exhibited a truer outline or better shape, and if she had come out a lit- more firmly in flesh she would have been very difficult to set aside for the championship." At Smithfield she improved on her position and beat the Short-horn-Aberdeen in a canter for the Cross Cup at the London Show.

A fine portrait of this heifer "Flora McDonald," appeared, April 2, in the London *Live-Stock Journal*.

2d. Sire Short-horn (Knickerbocker, the late Jas. Bruces' Burnside, prize bull); dam Polled roan.
3d. Short-horn-Galloway.

SPECIAL BUTCHERS' PRIZE.

1st. Sire Aberdeen (Paris 2d); dam Short-horn or pure Polled—uncertain; Black Poll; age, three years and ten months.
2d. Pure Short-horn; age, four years.

AT SMITHFIELD—YEARLING CLASS (STEERS).

1st. Same as first at Birmingham in same class; "carried four firsts this year."
2d. The fourth at Birmingham.

3d. Sire Short-horn bull; dam Polled blue-roan.
The third Birmingham winner was "left out in the cold."

TWO-YEAR-OLD CLASS (STEERS).

1st. Sire Aberdeen Poll; dam Short-horn; blue Polled. First also at Birmingham.

2d. Sire Short-horn; dam Polled; Black Poll.

3d. Sire Polled Aberdeen; dam Short-horn. Hereford; Black Poll.

(The Birmingham second was not ticketed here.)

THREE-YEAR-OLD CLASS (STEERS).

1st. The Birmingham first, which was here reserved for best ox (a white Short-horn getting cup for best ox). The *Live-Stock Journal* says: "We are inclined to think the cross-bred steer should have obtained the coveted position. By the large majority of on-lookers his success was expected, and the contest was certainly very close." In another part of the report the paper says: "We think he was clearly entitled to it."

2d. Sire Short-horn; dam ———(?)

3d. Sire Short-horn; dam ———(?)

HEIFER CLASS.

1st. The first-prize animal at Birmingham. Bearing in mind the remarks on the first-prize old steer, above, "which should have been best ox," this heifer must have been a remarkable one to have beaten that ox for the Cross Cup, which she did. Again the *Live-Stock Journal* says: "But for certain trifling defects she would have been an exceedingly stiff opponent to Mr. Stephenson's champion heifer."

2d. Sire Short-horn; dam Polled; Black Poll.

3d. Sire Short-horn; dam cross-bred Short-horn-Polled; red-and-white.

The *Mark Lane Express*, December 14, 1885, states: "The heaviest beast in the show was Mr. Wm. Tasker's ox, sire Aberdeen, dam by a Short-horn, bull from an Ayrshire and Short-horn cross cow; age 3 years, 8 months 1 week; weight 21 cwt. 3 qrs. 20 lbs."

ANOTHER LIST.

The following is a detailed list of principal prizes won by Aberdeen-Angus or their crosses.

NORWICH.

Cross bred.—Champion prize of 20 gs. for best beast: R. Loder, Short-horn sire, Aberdeen-Angus dam. President's prize, £20, to the same beast. Champion prize of £15 for the best cow or heifer in the yard: C. Stephenson, Aberdeen-Angus, pure-bred.

Steers of any breed, including cross or mixed breed, not exceding two years old: First prize, R. Wright (Lincoln), cross-bred, sire Polled Angus, dam Short-horn, bred by W. Curry (Hurworth), second prize, C. Stephenson, pure-bred Polled Aberdeen-Angus.

Steers of pure bred other than Short-horn or Red Polled, not less than two years and nor exceeding three years old: reserved and highly commended, W. B. Greenfield (Bedfordshire), Polled Aberdeen-Angus; commended, C. Stephenson (Newcastle), Polled Aberdeen-Angus.

Ox or steer of any pure breed other than Short-horn or Red Polled, not less than three and not exceeding four years old: first prize, G. Townshend (Soham), Polled Aberdeen-Angus, bred by J. W. Barclay, M. P., (Aberdeenshire).

Cross or mixed-bred steer, not less than two and not exceeding three years old: First prize, R. Loder, sire Short-horn, dam Polled Aberdeen-Angus; second prize, T. Freshney (Lincolnshire), sire Polled Aberdeen-Angus; dam Short-horn; reserve, Mrs. Meynell Ingram (Leeds), sire Short-horn, dam Polled Scot.

Cows of any breed other than Short-horn or Red Polled: First prize, C. Stephenson (Newcastle), pure-bred Aberdeen-Angus; second prize, W. B. Greenfield, sire Aberdeen-Angus, dam a cross-bred red cow.

BIRMINGHAM.

Elkington challenge cup, value £105, for the best animal bred and fed by the exhibitor: C. Stephenson, Aberdeen-Angus heifer; also President's prize and Scotch cup.

Cross-bred animals, special prize of £30: First prize, for oxen exceeding four years old, R. Loder, Short-horn sire, Aberdeen-Angus, dam; second prize, Messrs. T. and J. B. Freshney, sire Aberdeen-Angus, dam Short-horn.

Steers exceeding two and under three years old.— First prize, £20, R. Wright, sire Aberdeen-Angus, dam Short-horn.

Cross-bred cows or heifers—First prize, £15, W. B. Greenfield, Aberdeen-Angus sire, dam cross-bred.

Butcher's prize of £15—J. Cridlan, Aberdeen-Angus.

SMITHFIELD CLUB—LONDON.

Champion prize of £105—C. Stephenson, pure-bred Aberdeen-Angus heifer. Gold medal. Silver cup, value £50, for best heifer or cow, Scotch cup—C. Stephenson, pure-bred Aberdeen-Angus heifer.

THE BREED THAT BEATS THE RECORD. 135

Cross-bred steers, not exceeding two years—First prize, R. Wright, sire Aberdeen-Angus, dam Short-horn; third prize, O. C. Wallis, sire Short-horn, dam Aberdeen-Angus.

Cross-bred steers, above two years.—Second prize, O. C. Wallis, sire Short-horn, dam Aberdeen-Angus; third prize, C. Stephenson, sire Aberdeen-Angus, dam Short-horn.

Cross-bred steers, above three years—First prize, R. Loder, sire Short-horn, dam Aberdeen-Angus; third prize, W. Tasker, sire Aberdeen-Angus, dam by Short-horn.

Cross-bred heifers, not exceeding four years old—First prize, W. B. Greenfield, sire Aberdeen-Angus, dam first cross of Short-horn; second prize, J. Stephen, sire Short-horn, dam Aberdeen-Angus.

In addition to the above shows, the Aberdeen-Angus cross, and also the pure-bred, have figured largely in the prize lists at York, Leeds, Oakham, and other fat cattle shows.

THE RECORD IS OF "ABERDEEN-ANGUS" ONLY.

The above splendidly brilliant record for 1885 of the Scotch Polled was entirely made by the Aberdeen-Angus. This has been the case as long as we remember; Galloways being always absent.

The *Agricultural Gazette*, (London), December 13, 1880, reporting on the "Scotch Polled Breeds" at Smithfield Fat Stock Show, said: " Here the Galloway made no fight—the prizes all went to the Angus type." And the *Daily Free Press*, December 13, 1885, has the usual stereotyped phrase: " Though the class was open to Galloways, none put in an appearance,"

We may here, therefore, allude to the "difference" between the Aberdeen-Angus and the Galloway so often asked. Mr. Gordon, Chief Inspector of Live Stock for Queensland, writing in the *Queenslander*, 1883, a very high Scotch authority, says:

"Attempts have, on many occasions, been made to identify the Aberdeen-Angus with the Galloway. It it unnecessary here to traverse the ground so often traveled, to show that the two breeds have little in common beyond the color and the absence of horns. Suffice it to say that nothing in the history of either has been found to connect the two breeds. Mr. E. B. Woodhouse—who is a very careful observer in all matters connected with cattle—in his report on the late Royal Show in England, published in the *Queenslander*, notes the marked difference in the outward appearance of the two breeds. The sleek skin and elastic touch of the Aberdeen cattle is in the marked contrast to the shaggy coat, thicker hide, and curly neck of the Galloway. In the hind quarters the difference is also most marked, the Galloways being wanting in that full development which is characteristic of the improved Aberdeen cattle." Besides these, the "conspicuousness of the absence" of the Galloway from the natural and standard " platforms " of demonstration—Birmingham, Smithfield and other English Fat Stock shows, and Islington (London,) Fat cattle market.—is the chief and most signifficant " difference."

Part IV.

THE LONDON CHRISTMAS MARKET, 1885.

The following is from the report of the *National Live Stock Journal*: "Of even greater pecuniary importance to the British farmer than the shows at Birmingham and Islington, is the annual Christmas market of the metropolis. Here are to be found the very best commercial cattle exhibited during the year. The London market is always the choicest, and at Christmas time the selections are intended to supply the dinner tables of the most exacting judges of good beef. This year the market loses none of its high character. The quality taken all around has been extremely satisfactory. I do not think that the number of first-rate animals were so large as on former occasions, but from end to end the lots have rarely been better. Of course among the 7,400 cattle shown, there was a considerable 'tail' of really poor specimens, quite unfit for the trade, but about two-thirds of the muster were nearly all that could be desired. The preference of the London butchers is still very decidedly in favor of the black polls—and it is desirable to explain that by that term is meant the Aberdeen-Angus and their crosses, for the Galloways were not represented. Customers would not in the early morning, when the best trade was done, look at colored-horned animals so long as good hornless blacks were to be had. This partiality was never more strongly manifested. Farmers who had mixed lots soon got their polls cleared out, while they had to wait until the day was well advanced before they received any offers for bigger animals of varied shades. The polls have certainly acquired a most enviable reputation at the finest market in England, and it would be criminal folly if the breeders ever allowed it to sink. There is not much danger of their doing so, for they have the substantial inducement of immediate monitary reward to persevere in their efforts to keep their favorites in the front rank.

The supply was much too large for the demand. The Londoner is suffering from severe business depression, and is unable to purchase the choice joints that he likes to indulge in at Christmas time. Compared with last year, there was an increase of nearly 2,500 cattle on offer. Of that number fully 1,000 came from the northeastern counties of Scotland, the total supply from that country being 2,580. The top price for nicely finished cattle of about 7 cwt. to 8 cwt. was 5s. 8d. per stone of 8 lbs., but many good specimens made no more than 5s. 4d. per stone, while heavier weights did not fetch more than 4s. 10d. to 5s. 2d. The market has been a great disappointment to feeders. Last year an unexpectedly brisk business was experienced, and in the hope that a similar surprise was in store on this occasion, an exceptionally large number of beasts were retained. Business was very slow, and a clearance could not be effected. Among the animals shown I noticed two famous polled bulls—Prince of the Realm, a frequent winner at the national shows, and Wedgwood, for some time in Mr. Auld's herd in Aberdeen, shire, an exceedingly well-bred sire. Both these bulls were soon disposed of.

Apart from the formidable Scotch contingent there were a few very fine Herefords; there were also some creditable Devons, and a few heavy Welsh. The Shorthorns were also a very fine muster of large, well-formed cattle, and they, of course, contributed a good share in the production of the splendid crosses which formed, next to the Aberdeen-Angus, the main feature of the display.

The trade was the worst that has been experienced for sixteen years, last year's top price having been 6s. 2d. per stone, while in the two preceding years the highest quotation was 6s. 4d. Altogether the market has heen greviously disappointing in everything, except the quality of the stock exhibited."

This market is assuredly the very best market in the world for the choicest, primest and ripest cattle that the world can produce. It is the place where every breeder and feeder of whatever breed, who believes he handles that class of stock capable of standing a crucial test in respect to choiceness, ripeness, primeness and highest quality, has in his view, from the very beginning, to devote these animals coming up to such necessary perfection. We hear of "Polled Scots," "Prime Scots," and we hear of the "Aberdeen Polls" and their crosses, as we might necessarily expect, in terms of the highest praise in all the above respects. But we never hear of any other Scotch Poll in such laudatory terms at all. Not being able to stand the test they are not forwarded, being disposed of at the "side" markets. Nowhere else as in London, as the world well knows, is the very best beef in demand and required; and nowhere but there can the best so invaribly find its best market. "It should be particularly remarked that when an English writer speaks of Polled Scots he means the Aberdeen-Angus which alone he is accustomed to see at London and Birmingham, for a Galloway is a *rara avis* there. This should be made plain once for all to breeders on this side. Take up any of the British journals and we find that when they write on Polled Scots or Prime Polled Scots or Prime Scots they invariably and pointedly mean 'Aberdeen.' In London, at Smithfield shows and at the Islington Great Market we hear of nothing but prime Aberdeenshire as descriptive of those Prime Scots and Prime Polled Scots. It is Aberdeens Polls that are the Prime Scots or Prime Polled Scots. In every well informed quarter this is well understood, any other allegation is a perversion of fact and absolutely devoid of the basis of substantiability." So writes a correspondent in the *Kansas City Live Stock Record*.

Another in the *Breeder's Gazette*: "I would here confirm what a Scottish correspondent says about the identity of the Aberdeen-Angus with 'Polled Scots' and 'Prime Scots' in England. Take up the *Agricultural Gazette, Live Stock Journal, Field*, etc., and you will see these terms used to indicate the Aberdeens and no other. All the great achievement of the Aberdeens in France and England are in all the newspapers and heralded as won by the 'Polled Scots.' We must therefore be particular about that. The *Live-Stock Journal*—to which we have to resort for most of our live-stock information—now annually gives prominence to a capital report of the London Christmas market. Its report for the year (1885) says: 'As regards quality, the average meat was never higher. As in former years the Scotch cattle were decidedly the best. A large proportion of them were black polled, either pure-bred Aberdeen-Angus or crosses between that breed and the Short-horn the character of the polled predominating."

This is the *Breeder's Gazette* report: " The Scotch beeves were excellent in character and quality, and gave a maximum amount of meat on very little bone. Indeed, the lot were very even and well developed ; and showed more or less of the polled characteristics, although some were the result of a cross between the polled Aberdeen and Short-horn, or *vice versa*. The Herefords and Short-horns also showed up well, but the whitefaces predominated over the 'red, white and roan.' While the Devons and Welsh runts may be next taken in the scale of good quality.

" The home-bred cattle from Norwich and Suffolk were Red Polls of prime value, and the Short-horns from various districts were not without merit, but at the great Islington market the Black Poll, West Highlander, and Hereford have pride of place in the estimation of butchers and their customers."

And that of the *Kansas City Live Stock Record*: " The principal breed was the Polled Aberdeens with a fair sprinkling of West Highlanders. These obtained the top price of the market, namdly, $1.36 per 8 lbs.''

Same paper, March 25, 1886, on same market: " The best Scotch cattle such as the Aberdeen Polls."

The *Mark Lane Express*, Decebmer 14, 1885, reports: " In point of quality and condition to-day's display of stock may be looked upon as distinctly satisfactory, and the number also was good. No falling off could be reported in the Scotch arrivals. It is always expected that the Scotch breeders will keep up their reputation for excellent stock, and no surprise s therefore felt that to-day proves no exception to the rule. It would be curious were a falling off noticed. Scotland retains her traditional position. Cross-breds were, as usual, a numerous class, and were second to none regarded in the light of a profitable vehicle for supplying the principal markets with prime meat. Taken altogether, the show may be described as a success, and quite up to the average. At to-day's market, however, the exhibition was a very distinct improvement over the previous two years both in point of number and condition. The average rate per head was, we should say, distinctly heavier. The Scotch were a superb show, and all the best known breeders were, as usual, well represented. They certainly formed the most attractive portions of the market. Not that there was nothing else to attract attention. Herefords, Sussex, and Welsh Runts were all well represented. In fact, the market all around was a decided success."

ADDENDA.

A LONDON POLLED ABERDEEN SOCIETY.

The daily *Free Press*, Aberdeen, Scotland, December 7, 1885, said: " It will be interesting to Scotch exhibitors to know that there has been established in connection with the Smithfield Club, in London, a Polled Breeders' and Feeders' Annual Dinner Society. The first meeting has been held. Mr. R. Walker, factor, Altyre, who has been one of the most successful exhibitors of Scotch cattle in London, has been unanimously appointed President of the Society, and Mr. Clement Stephenson, Sandyford Villa, Newcastle, has been appointed Vice-President. The object of the Society is specially to bring exhibitors of the polled cattle together in London during the Smithfield week, and to further the interests of the breed of polled cattle, which are making remarkable progress in England. The membership includes several prominent Scotch and English exhibitors of polled and polled cross-bred cattle."

This was instituted before the Smithfield victories of the year.

THE BREED EXTENDS ITS OUTPOSTS MORE AND MORE.

The Banffshire *Journal*, in its reports of the late Birmingham and Smithfield shows, notes as the consequent result of the unparalleled achievements of the Aberdeens at these exhibitions the further extension of the breed into England by the establishment of new herds. This is very gratifying. It is remarkable, assuredly, the great number of herds of Aberdeens in England. There are more herds of Angus in England—outside its native limits—than there are Short-horns (or any other English breed) in Scotland. Take as a test the Breeders' Directory in *The Live-Stock Journal Almanac*. In this we find only two Short-horn herds in Scotland mentioned, while there are no fewer than five Aberdeen-Angus in England.

"Major Irwin, Lynthow, Carlisle, who has been breeding Short-horns, is to replace them by a herd of Polled Aberdeen-Angus, which he is now forming."

Such items we are constantly reading now in our home papers; showing the increasing favor in which they are regarded in England.

We could give very many more similar "modern instances" in Scotland, England and Ireland where Short-horn breeders have sold off their favorites and gone in for the "World-Beater" breed.

In Vol. X of The Polled Herd Book, more entries are made from England than from America. American breeders of course naturally now prefer to record in the American Register. The number of English adherents in that Vol. is 20, and still there will be more breeders in the future. Its most conspicuous convert in England from the old faith, is Clement Stephenson. There are also other very prominent Short-horn breeders now breeding Aberdeens contemporaneously with the red, white and roan. In America, the most prominent adherent may be truly said to be Hon. M. H. Cochrane, of "Hillhurst" fame ; who, from breeding four-thousand-guinea Duchesses, has gone in for the "coming" Aberdeens of which he is making a "specialty." We trust to see him producing some "thousand-guinea" Ericas and Prides. Col. G. W. Henry, of Angus Park, Kansas City, Mo., another champion "against all" is also a convert. So is Mr. T. W. Harvey, who, has secured Mr. Wm. Watson, as his newly appointed director of the destinies of the Turlington Doddies; "Willie" Watson, Angus foes are, by the *Breeder's Gazette*, cautioned to look out for—"as he is to Clement Stephenson, the Herefords and Short-horns next fall."

Such tests as these, as Youat said of the Collings' sale—so unmistakably demonstrated by the public estimate as declared by the results of the auctioneer's hammer, can't be overlooked. Such are the foregoing and the following from that "live" organ, the *Kansas City Live-Stock Indicator:* "A Kansan in Auld Scotia: The *Banffshire Journal*, of Scotland, dated February 9th, has the following mention of a wide-awake Kansas man who has been rummaging around among some of the best of the polled herds to be found in their native country: 'In a notice in *The Field* of Saturday last of the herd of Aberdeen-Angus cattle belonging to Mr. C. Stephenson, Newcastle (the breeder of Luxury, the heifer which won the champion prize at the last Smithfield show), we find the following—' Judge Goodwin of Beloit, whose visit to Mr. Stephenson jumped well with our own, told us that on the Goodwin Park stock farm, worked by a brother and himself, they had tested on a small scale the relative and comparative merits of Short-horns, Herefords,· Angus, Galloways, Jersey and Holstein cattle, side by side on the natural pastures, without any additional food, and the result was that the Short-horns were first drafted, next two dairy breeds, then Herefords, and lastly Galloways. The evidence in favor of the Angus cattle was overwhelming, especially as regards adaptibility to climatic conditions, hardiness of constitution, and ability to thrive on little food; indeed, when once in condition, it would be difficult to starve them. This firm has now 125 head, including some very choice animals; indeed, they owned Judge, of the Jilt family, which was the champion bull at the Paris show in 1878. Messrs. Goodwin breed partly for the ranches, but principally to sell bulls to small farmers in Kansas, to improve the common stock of the country, and had an increasing demand."

These are the best "comparative" tests. Who cares a jot for the so-called scientific college comparative experiments that have been published? We know that at a certain Agricultural College from which emanate voluminous periodicals, called comparative "tests" made on the beef and milk breeds" the students graduate with the dogma *e.g.* that the Aberdeen-Angus is no "feeder," and also come away with "a prejudice against the Aberdeen-Angus." What is the public opinion, however, of these "tests?" We only allude to the half-breed experiments.—Here is what the *Mark Lane Express* has recorded: " The experimental trials in connection with the feeding of cattle and sheep as conducted by ―――― of the ―――― Agricultural College, ――――, come before the public in a voluminous report printed by order of the Legislative Assembly. The tabulative details are positively bewildering, and when the number of amimals experimented on is taken into consideration the game seems scarcely worth the candle. * * * * *

"The elaborate analyses * * * * do not seem very tangible. With regard to the experiments with sheep, they are still more intangible in their nature and rendering, and altogether the Professor seems to have shown a *mountain of labor with a mouse for result.*"

The test of the *dollar* is not the least important. The general average of prices realized at public sales of Short-horns, Herefords and Aberdeen-Angus for 1883, 1884 and 1885 in America, have been in favor of the last named breed.

And still the breed is ahead at sales. The following is the report of the "last" sale, that of Hon. M. H. Cochrane, at Chicago, May 16, from *Breeder's Grzette:* "The draft of doddies included some yearling and two-year-old heifers of exceptional merit, and a number of good cows with calves at foot or forward in calf to Paris 1163. The most conspicuous feature of the sale was the absence of Angus breeders, there being but very few in attendance. With the exception of two Lady Idas, a Favorite, a Baromess, and an Emily of Kinochtry, the lots were not fashionably bred, but were for the most part from standard and well-established herds, but of such excellent individual merit were they that bidders seemed inclined to follow them well to their worth. The men who purchased were chiefly beginners or small breeders who appeared to appreciate the character of the cattle offered them. The crowd was thoroughly tired and restless by the time the blacks were reached, and to add to the coufusion, a heavy storm broke with such force as to almost bring the sale to a stop on account of the darkness and the din. In spite of these disadvantages an average of $371.65 was scored, and had not a cow which had aborted a few days before the sale been included, the average would have reached $380. The magnificent Lady Ida heifer Lady Lyra Hillhurst 3906 was the highest priced lot, going to Elias Trumbo, Ottawa, Ill., for $800. The best bargain of the day was Lord Lyons, a full brother to Lady Lyra Hillhurst, and pronounced by Mr. Cochrane the best individual of any breed he had ever bred at Hillhurst. This unusual youngster went at the low figure of $450 to Ben F. Elbert, Albia, Ia., who is quietly building up a herd of 'top' Angus doddies." This certainly beats the record as to prices, this season too. For the sale of the Attrill Duchess Short-horns, next day at same place, cannot be put against this sale. When Cochrane sells some of his Prides and Ericas, he will leap into the thousands; as he did a few years ago, getting over $3,000 for one heifer at public sale.

THE POLL TAKES THE PRIZE IN A KEEN "MILK-COW" COMPETITION.

The British correspondent of the Kansas City *Live-Stock Record*, March 18, 1886, writes: "An interesting competition was held at Edinburgh last week by Messrs. Oliver & Son, live-stock salesmen in the southeast of Scotland. They offered $200 in three classes for milk cows; and there was a good entry of useful dairy cattle in all three. The first class was for Short-horn or cross-bred cows, and two dozen faced the judges, premium honors being gained by a Black Poll. Sixteen Ayrshires were entered for competition, and a dozen in the open class for any breed. The competition was very keen; and at the sale which followed the dairy stock entered were disposed of at better rates than have been current for some time."

CROSSING.

From *Live-Stock Journal*, February 26: " Mr. Dickinson, Roos, Hull, has sold two Aberdeen-Angus bulls *for crossing purposes* to go to the west of Ireland. Mr. Robert Bruce, Great Smeaton, Northallerton, has purchased three Aberdeen bulls for farmers in the north of England *for crossing purposes*. The Aberdeen-Angus bulls put to Short-horn cows give a good account of themselves. There are quite a number of farmers in the northeastern counties of England using Aberdeen-Angus bulls."

Mr. James A. Cochrane, of Hillhurst, Que., writes (*Breeder's Gazette*, March 25, 1886,) of his recent trip to Britain: " Before leaving the other side, I visited some of the largest and best herds of Polls in the South. The interest in polled cattle is rapidly increasing. Bulls of the breed being bought by influential men in all parts of the country for crossing purposes, largely in consequence of the sweeping victories of Aberdeen-Angus cattle and their crosses at the last Smithfield show." Judge J. S. Goodwin, of Beloit, Kansas, a large breeder of Aberdeen-Angus, reports similarly to us on his recent trip to the old world.

The following appears in the April (1886) number of the *Canadian Live-Stock Journal*: " The records of the Smithfield Club, (London,) have shown that for the joint purpose of providing beef and coming early to maturity, no race of cattle can beat the Polled Aberdeen-Angus. * * * * a breed of cattle that in many important respects outstrips all rivals. * * * * as a breed their type is almost more firmly fixed than any other race of cattle. *No variety exercises so strong an influence in the moulding of their progency, not even the Short-horn or Hereford or Devon.* That has been a distinctive feature of the Polled Aberdeen-Angus cattle from the first."

A correspondent of the London *Live-Stock Journal*, April 22, 1886, says: "I have great faith in the cross between a Polled Angus bull and Short-horn cow. I think for butchers' purposes there is nothing can beat them. They make good weights, and for early maturity and lightness of offal they are second to none. They are also very good dairy cows. I question very much if any of the Herefords or Short-horns did more in one year than the Black Polls and Black crosses did last year, for in all the leading shows in England every first prize in the cross-bred classes was taken by this cross, and, with the exception of Mr. Loder's aged ox, whose sire was a Short-horn, all the others were sired by Polled Angus bulls, which I believe to be the most impressive sire of the day; and as there are a good many young sires of this breed all over England now, I hope to see blacks and blue-grey crosses winning again next year."

EARLY MATURITY.

George T. Turner, writing to the *National Live-Stock Journal*, February, 1882, said of the Aberdeens: "Their fine bone, thick flesh and hardiness are greatly in their favor as beef cattle; and the records of the London Show go to prove that they can lay claim to the front rank in respect of early maturity. In the annual tables published by the *Mark Lane Express*, the Scotch Polled cattle came out in a manner which is something of a surprise to English breeders, and they are in a fair way to come into more general favor with English feeders. For shop use, the Polled animal, on account of his smaller bone was a better cutting beast."

CHICAGO DEALERS' AND BUTCHER'S OPININIONS.

Harrison Miller writes *Breeder's Gazette*, April 29: "Regarded as beef-producers, they are not excelled, being compact and symmetrical, well fleshed, and their meat being of a most excellent quality. I remember helping my father to slaughter one a few years ago, and when we came to eat the beef, we all thought it was the best we had ever tasted, it was so sweet and juicy."

Another (English) butcher describing the slaughter of a heifer said: "Killed, she dressed a remarkable carcass. She was one of the best fed beasts ever seen, with so little internal room that *it was a surprise where the viscera had been packed*, and with plenty of lean meat, very fine bone, and showing remarkably small offal throughout."

The following is from the Chicago *Drover's Journal*, April 29: "S. D. Seaver, of Mantino, Ill., is a farmer and a practical stockman. He breeds Short-horn and Polled-Angus cattle, not for breeding purposes, but for beef. He was among the visitors at the yards and brought in three cars of cattle, all of his own feeding, consisting of half-blood Polled-Angus steers, pure Short-horn steers, a car of choice cows and heifers, and some medium beef steers. The Short-horns were very pretty and attracted not a little admiration, but the attention chiefly centered about the pen containing the 16 grade Angus. They were out of a pure-bred

Angus bull and Short-horn cows, and were entirely hornless and almost uniformly black. The majority of them were just two years old this month, but some in the lot were coming twos in July. The two-year-old Short-horns sold at $5.00; the medium steers sold at $4.55, and the heifers at $4.12½, but at a late hour the black cattle had not been sold." Next week it said: "The Polled-Angus steers, 16 head, averaged 1224 lbs., sold at $5.40. They were bred and raised by Mr. S. D. Seaver, of Mantino, Ill. The cattle were carrying two-year-olds in April to July."

"IN THE WEST."

In connection with the notice of the Victoria Ranch, the following from Mr. Thomas R. Clark, is from the *National Live-Stock Journal*, July, 1880: "On reading the article 'Polled Angus Cattle in Missouri,' in the *Journal* for June, quoting a letter to the Kansas City *Price Current*, written by Mr. Joseph H. Rea, of Carrol county, Mo., giving his opinion and experience with a lot of young steers from my stock farm in Kansas, of this breed, I deem it proper to state, for the information of your readers, that the animals he refers to were a lot of two-year-old steers, the first cross from pure black polled bulls on the common Texas cows, sent by me last fall to the Kansas City market. The above statement shows to be a very important factor to be taken into consideration in forming an opinion of the merits of any breed of cattle. It is very gratifying to me to know that Mr. Rea has such a high opinion of these animals, and it convinces me that two or three moves from the common Texas stock will render them still more appreciated. I believe them eminently adapted to the Western Plains, by reason of their extreme hardihood—thriving where others would loose flesh—the entire absence of horns, and above all, the superior quality of their top price in the English markets. The late Mr. Grant and myself imported four of these young bulls, in 1873, and it is from the progeny of these three bulls, (probably the only ones then in the United States,) that I have received so many compliments on the style and quality of my cattle from so many quarters. I have an abiding faith that these Polled Aberdeen cattle are yet to become very popular in the United States, possessing so many very valuable qualities.

Mr. Wm. Watson, superintendent at Turlington, writes me: "The Western men have fairly set their hearts on the Aberdeen-Angus. Metcalf the great cattle man had some pure Short-horn cows in calf and with calves at their feet to a Black Polled bull. Four of the cows strayed away during the first storm this winter. They had suckling calves with them. The herdsmen went after them and came upon them some twenty miles from the ranch. The four Stort-horn cows were all lying dead. The four black calves were standing beside their dead mothers, but were strong and hearty and were able to walk back to the ranch along with some others. These calves were only about 12 weeks old, on an average. So you see what the black cross effects. Governor Routt, who has his herd near Denver, says his Aberdeen-Angus and half-breed calves were all saved, although dropped in the snow. All the other sorts nearly all died. These are strong arguments for the Aberdeen-Angus."

The Drover's Journal, March 25, 1886, noting the arrival at Littleton, Mass., of a large importation of Aberdeen Polled cattle by Messrs. Kirby & Cree, of

Fort Stanton, New Mexico, says: "The cattle were purchased with great judgment and care by Andrew Mackenzie, of Dalmore. Ross-shire, and selected by him personally during the past four months. The name of the ranch for which these cattle are destined is called the 'Angus VV. Ranch,' situated in Lincoln county, New Mexico. The property is of large extent, embracing a large area of rolling country, watered by the Rio Doso, Rio Bouito, Rio Saludo, Little Creek and Eagle, five rivers that afford a large supply of water. In the pastures at present are extensive herds of high grade Short-horn and Hereford cattle. The owners were desirous for the most likely *corrective* of the tendency in these descriptions of stock to develop bone, and their inquiries resulted in the conviction that the Aberdeen-Angus Polled cattle, were the most likely to impart the qualities of *low standing, thick flesh and early maturity.* The bulls, with one exception, will be placed on the ranch, in company with their Short-horn and Hereford grades, with which they will be crossed."

ABERDEENS PREFERRED TO THE JERSEYS.

Mr. C. R. C. Dye, well known as a breeder of Jerseys at Troy, O., has made a new departure, as will be seen from the following communication to the *Breeder's Gazette*, May 20, written by him under date of April 25 from Scotland: "After my disastrous fire and destruction of the bulk of my Jerseys, of which I wrote you, I determined to come abroad and replenish my herd, with little or no idea of buying 'black doddies,' but I thought I would take a look at them before going to the Island of Jersey, and the consequence was that I was so pleased with them that I have purchased as a foundation for a herd sixty-five head—sixty females and five bulls."

THE BREED THAT BEATS THE RECORD. 157

I may note here that some remarkable cattle transactions have taken place in Scotland, by Mr. Geary, of Canada and others. This gentleman has purchased the two whole herds of Gavenwood and Glenbarry formerly owned respectively by Mr. John Hannay and Mr. J. W. Taylor. The two herds consist of nearly 100 head. These transactions taken in conjunction with that regarding Justice, are surely grand testimony of the new world in favor of Old Scotia's National Breed.

CONCLUSION.

"THE WORLD'S BEEFER."

The following declaration is by Professor Brown, principal of the Guelph Agricultural College, Ontario, as to the world's new beefer which "is unquestionably the Aberdeen-Angus Poll. In these times of specialties this breed of cattle is bound to fill a big place in the world's products. The hardiness. early maturity, general quality and weight of the Poll can not fail to lead. It was really a very pleasant duty to inspect, as I did, nearly every prominent herd of these in Scotland, and to see so much 'canny,' forseeing, practical judgment exercised in their extensive production. I could not buy from some at even £1,000 a head, and yet I gave the highest price that had ever been paid for a bull. The black diamonds of the north of Scotland will make warm ground for the Short-horn and Hereford."

"THE COMING STEER."

Mr. J. J. Hallet, in an address before the Carroll county (Ill.) Breeders' Association, reported in the *Breeder's Gazette*, January 6, 1886, indicates how we are to obtain " The Coming Steer:" "The almost complete success that has been accomplished in the line of breeding at will, and the genuine improvement in every way upon the steer, both as to quantity and quality, seems to be impossible in the minds of some otherwise substantial farmers. And if there should be any such present, I will state for their encouragement that the question can no longer be a doubtful one when breeders have been able to breed horns on or off their steers' heads just as they wish it to be. There is a breeder now living in Carroll county, a prominent member of this live stock association in good and regular standing in the society, who, after completing his arrangements with an imported colored gentleman from Canada, by the name of Aberdeen P. Angus, publicly challenged his entire herd of cows, who carry horns of every conceivable size and shape, to produce a single calf that will ever grow a horn upon his own head, *and about three years have passed without a single horn to be seen.* If a vote should be taken here to-day upon the question of what the coming steer will be, there would be at least one vote for the muley."

This is doing good. The editor of the *Drover's Journal* comments thus: "Apropos of the late agitation about dehorning cattle, it is stated that three years ago an Illinois stockman introduced a polled Aberdeen-Angus bull into his his horned dairy herd, and, as he expresses it, 'publicly challenged his cows to produce a single calf that should ever grow a horn upon its own head,' and up to the present time they have not been able to do it. The *New England Farmer*, whose editor has had satisfactory experience in the same direction, commends this *breeding off a worse than useless excresence*, and hopes there will be a general effort to the same end, which, he says, does not involve special sacrifice of other peculiarities and good qualities of favorite animals."

Many have the stupid whim that horns are ornamental. We are quite surprised to see that advanced journal, *Farmer's Review*, March 3, 1886, saying so; though it admits the great harm they are capable of. But the *Drover's Journal* says, right out: "*Horns are certainly not useful on beef animals, and not very ornamental.*" The horns must go. We do not consider this a *great* argument to advance the interests of the prepotent, pre-eminent, Polled Aberdeen-Angus. Indeed, the *National Live-Stock Journal*, even on the first appearance of the breed in this country said: " The want of horns is, in our opinion, but *the least* of their good qualities."

"MORALLY CERTAIN TO WIN."

The *Breeder's Gazette* observes that "the admirers of the Angus in particular are showing a disposition to make a strong pull for popular favor. The record their favorites have made at the fat-stock shows both at home and abroad has served to nerve the already strong hands in which the breed is largely held in America, and there is every prospect of all reasonable prosperity for the 'doddies' during the coming year. Home-bred polled bullocks will soon be an established feature of our fat-stock shows, and there is every reason for assuming that the success which they are morally certain to win both on foot and upon the block at these exhibitions will stimulate the demand for pure-bred specimens for stock purposes." And echo answers: "Few as their numbers have been, they have already made their mark in the carcass ring in a way that warns competitors to beware."—*National Live-Stock Journal.*

The *American Agriculturalist* (New York), August, 1885, thus declares itself in favor of "Angus Doddies:" "Among all the excellent breeds of neat cattle for which the world, and the 'new world' particularly, is indebted to the British Isles, none take a higher economical rank than the Angus 'Doddies,' which have become great favorites with the graziers throughout our great feeding region. They are a compact, well-shaped, hardy race, of tremendous potency of blood. Their coats, which are fine and heavy, afford all needed protection against rain and cold, wind and hail, to all of which the herds of the plains are more or less subject. * * * * * *

"There is no better beef in the world than that of the Angus cattle. The choice parts are well developed, the fat is laid on evenly, and the meat beautifully marbled, while the percentage of offal is very small. *
* * * * * *

"The future of the breed in this country is assured. It is admirably adapted to shipping by rail or steamer. Its grades possesses the characteristics of their sires to a remarkable degree, maturing early, keeping themselves in thrifty growing order on only fair forage, and fatting up very quickly when they get grain. They have won not only prizes, but public confidence, and it is hardly too much to say (though perhaps it is not modest in us), that the highest prize they can win seems now within their grasp—*the confidence of the American farmers and cattlemen, that they are the best breed of the beef cattle in the world.* The *American Agriculturist* has long (and persistently) advocated Polled cattle for the plains, and especially for shipping alive to European markets. There is nothing equal to them for this end, and the facts as reported by shippers and breeders abundantly corroborate these views."

The French journal *Le Fermier*, has declared that: "The Angus is one of the most remarkable breeds of cattle, and is bred under conditions and by means which make it worth more than any other of the attention of rearers."

The following is the general grand summary, brought to a focus and common percentage—as recorded in the official report of the Seventh Annual American Fat Stock Show. That the extract is a fitting conclusion to our labors, will be admitted: " It will be seen from the above awards of premiums by expert judges at the seven shows, if we take into consideration the number of entries of each class of cattle shown, that prestige is as follows:

" First, Aberdeen-Angus (it having received 219.15 per cent. above its proportional amount); second, Grade Angus, (it having received 192.4 per cent. above its proportional amount); third, Short-horn, (it having received 46.21 per cent. above its proportional amount); and fourth, Herefords, (it having received 22.21 per cent. above its proportionate amount."

To apply a well known and very appropriate phrase, the author may be permitted " lastly " to declare: *They are the chunky sort that means business.*

BIBLIOGRAPHY.

The following works ought to be consulted by Aberdeen-Angus men: " HISTORY OF THE HIGHLAND

AND AGRICULTURAL SOCIETY." by Alexander Ramsay, editor of the *Banffshire Journal*, editor of The Polled Herd Book, published 1879, price 16s.—HISTORY OF THE POLLED ABERDEEN-ANGUS BREED OF CATTLE." By Jas. McDonald, author of " Food from the Far West," etc., etc., editor of *The Live-Stock Journal*, etc., and Jas. Sinclair, associate-editor of do. Price 10s. 9d.—" CATTLE AND CATTLE BREEDERS." Fourth edition, just published, by William McCombie of Tillyfour, M. P., with a memoir of the author by Jas. McDonald, editor of the *Live-Stock Journal*, published 1886. Price, 3s. 6d.—The above are all published by Blackwood & Sons., George St., Edinburgh, Scotland, and a postal note for $10 will deliver the three.

THE AMERICAN ABERDEEN-ANGUS ASSOCIATION.

The address of the Secretary of the above Association, who publish the American Polled Herd Book, is: Charles Gudgell, Esq., Independence, Mo.

ERRATA.

The author is responsible for these two oversights: Page 4, line 10 from *top* for " Eccossais " read *Ecosse*. Page 89, line 4, from *top*, for " Journal " read *Indicator*.

Some other typographical errors being of a trifling nature need not be referred to.

" JUDGE'S " DIMENSIONS.

On page 20, the dimensions of the bull "Judge" have been incorrectly rendered by the printer. From Mr. W.R.Goodwin,Jr., I am informed "Judges's" dimensions were as follows: Height at shoulder, 4 feet 8 inches; heighth at hip, 4 feet 9 inches; girth, 7 feet 10 inches; length from top of poll, 8 feet 1 inch; weight, 2,600 pounds.

FINIS.

www.ingramcontent.com/pod-product-compliance
Lightning Source LLC
Chambersburg PA
CBHW031444160426
43195CB00010BB/846